Cristian V. Ciobanu,
Cai-Zhuang Wang,
and Kai-Ming Ho

**Atomic Structure Prediction
of Nanostructures, Clusters
and Surfaces**

Related Titles

Werner, W. S. M. (ed.)

Characterization of Surfaces and Nanostructures

Academic and Industrial Applications

2008

ISBN: 978-3-527-31760-8

Eftekhari, A. (ed.)

Nanostructured Materials in Electrochemistry

2008

ISBN: 978-3-527-31876-6

Vedmedenko, E.

Competing Interactions and Patterns in Nanoworld

2007

ISBN: 978-3-527-40484-1

Wilkening, G., Koenders, L.

Nanoscale Calibration Standards and Methods

Dimensional and Related Measurements in the Micro- and Nanometer Range

2005

ISBN: 978-3-527-40502-2

Reich, S., Thomsen, C., Maultzsch, J.

Carbon Nanotubes

Basic Concepts and Physical Properties

2004

ISBN: 978-3-527-40386-8

Cristian V. Ciobanu, Cai-Zhuang Wang,
and Kai-Ming Ho

Atomic Structure Prediction of Nanostructures, Clusters and Surfaces

WILEY-VCH Verlag GmbH & Co. KGaA

The Authors

Dr. Cristian V. Ciobanu
Colorado School of Mines
Division of Engineering
1610 Illinois Street
Golden Colorado 80401
USA

Dr. Cai-Zhuang Wang
Iowa State University
Ames Lab/Dept of Physics
12 Physics Hall
Ames, Iowa 50011
USA

Dr. Kai-Ming Ho
Iowa State University
Ames Lab/Dept of Physics
12 Physics Hall
Ames, Iowa 50011
USA

All books published by **Wiley-VCH** are carefully produced. Nevertheless, authors, editors, and publisher do not warrant the information contained in these books, including this book, to be free of errors. Readers are advised to keep in mind that statements, data, illustrations, procedural details or other items may inadvertently be inaccurate.

Library of Congress Card No.: applied for

British Library Cataloguing-in-Publication Data
A catalogue record for this book is available from the British Library.

Bibliographic information published by the Deutsche Nationalbibliothek
The Deutsche Nationalbibliothek lists this publication in the Deutsche Nationalbibliografie; detailed bibliographic data are available on the Internet at <http://dnb.d-nb.de>.

© 2013 Wiley-VCH Verlag GmbH & Co. KGaA, Boschstr. 12, 69469 Weinheim, Germany

All rights reserved (including those of translation into other languages). No part of this book may be reproduced in any form – by photoprinting, microfilm, or any other means – nor transmitted or translated into a machine language without written permission from the publishers. Registered names, trademarks, etc. used in this book, even when not specifically marked as such, are not to be considered unprotected by law.

Print ISBN: 978-3-527-40902-0
ePDF ISBN: 978-3-527-65505-2
ePub ISBN: 978-3-527-65504-5
mobi ISBN: 978-3-527-65503-8
oBook ISBN: 978-3-527-65502-1

Cover Design Adam-Design, Weinheim
Typesetting Thomson Digital, Noida, India
Printing and Binding Markono Print Media Pte Ltd, Singapore

Printed in Singapore
Printed on acid-free paper

Contents

Preface *IX*

1   **The Challenge of Predicting Atomic Structure** *1*
1.1   Evolution: Reality and Algorithms *2*
1.2   Brief Historical Perspective *4*
1.3   Scope and Organization of This Book *6*
         References *7*

2   **The Genetic Algorithm in Real-Space Representation** *11*
2.1   Structure Determination Problems *12*
2.1.1   Cluster Structure *12*
2.1.2   Crystal Structure Prediction *16*
2.1.3   Surface Reconstructions *19*
2.1.4   Range of Applications *21*
2.2   General Procedure *23*
2.3   Selection of Parent Structures *24*
2.4   Crossover Operations *26*
2.4.1   Cut-and-Splice Crossover in Real Space *27*
2.4.2   Crossovers and Periodic Boundary Conditions *28*
2.5   Mutations *30*
2.5.1   Zero-Penalty Mutations *31*
2.5.2   Regular Mutations *31*
2.6   Updating the Genetic Pool: Survival of the Fittest *33*
2.7   Stopping Criteria and Subsequent Analysis *34*
         References *35*

3   **Crystal Structure Prediction** *37*
3.1   Complexity of the Energy Landscape *38*
3.2   Improving the Efficiency of GA *40*
3.3   Interaction Models *41*
3.3.1   Classical Potentials *41*
3.3.2   *Ab Initio* Methods *42*
3.3.3   Adaptive Classical Potentials *42*

3.4	Creating the Generation-Zero Structures	44
3.5	Assessing Structural Diversity of the Pool	45
3.5.1	Fingerprint Functions	45
3.5.2	General Features of the PES	47
3.6	Variable Composition	48
3.7	Examples	51
3.7.1	Identification of Post-Pyrite Phase Transitions	51
3.7.1.1	Computational Details	52
3.7.1.2	Results and Discussion	52
3.7.2	Ultrahigh-Pressure Phases of Ice	57
3.7.2.1	Computational Details	58
3.7.2.2	Results and Discussion	59
3.7.3	Structure and Magnetic Properties of Fe–Co Alloys	63
3.7.3.1	Computational Details	63
3.7.3.2	Results and Discussion	64
	References	67
4	**Optimization of Atomic Clusters**	**71**
4.1	Alloys, Oxides, and Other Cluster Materials	71
4.2	Optimization of Substrate-Supported Clusters via GA	73
4.3	GA Solution to the Thomson Problem	81
	References	85
5	**Atomic Structure of Surfaces, Interfaces, and Nanowires**	**87**
5.1	Reconstruction of Semiconductor Surfaces as a Problem of Global Optimization	88
5.1.1	The Genetic Algorithm for Surface Reconstructions: the Case of Si(105)	89
5.1.1.1	Computational Details for Si(105)	89
5.1.1.2	Results for Si(105)	91
5.1.2	New Reconstructions for a Related Surface, Si(103)	95
5.1.3	Model Reconstructions for Si(337), an Unstable Surface: GA Followed by DFT Relaxations	99
5.1.3.1	Results for Si(337) Models	101
5.1.3.2	Discussion	106
5.1.4	Atomic Structure of Steps on High-Index Surfaces	107
5.1.4.1	Supercell Geometry and Algorithm Details	107
5.1.4.2	Results for Step Structures on Si(114)	110
5.2	Genetic Algorithm for Interface Structures	114
5.2.1	GA for Grain Boundary Structure Optimization	115
5.2.2	Structures Generated by GA	116
5.2.3	Grain Boundary Energy Calculations	121
5.3	Nanowire and Nanotube Structures via GA Optimization	123
5.3.1	Passivated Silicon Nanowires	123
5.3.2	One-Dimensional Nanostructures under Radial Confinement	130

5.3.2.1	Introduction *131*
5.3.2.2	Description of the Algorithm *132*
5.3.2.3	Results for Prototype Nanotubes *135*
5.3.2.4	Discussion *139*
5.3.2.5	Concluding Remarks *144*
	References *144*

6	**Other Methodologies for Investigating Atomic Structure** *149*
6.1	Parallel Tempering Monte Carlo Annealing *151*
6.1.1	General Considerations *151*
6.1.2	Advantages of the Parallel Tempering Algorithm as a Global Optimizer *153*
6.1.3	Description of the Algorithm *154*
6.2	Basin Hopping Monte Carlo *158*
6.3	Optimization via Minima Hopping *160*
6.4	The Metadynamics Approach *163*
6.5	Comparative Studies between GA and Other Structural Optimization Techniques *165*
6.5.1	Reconstructions of Si(114): Comparison between GA and PTMC *165*
6.5.1.1	PTMC Results *166*
6.5.1.2	GA Results *167*
6.5.1.3	DFT Calculations *167*
6.5.1.4	Structural Models for Si(114) *169*
6.5.1.5	Discussion *174*
6.5.1.6	Concluding Remarks *175*
6.5.2	Crystal Structure Prediction: Comparison between GA and MH *175*
6.5.2.1	GA Applied to Al_xSc_{1-x} Alloys *176*
6.5.2.2	Boron *180*
6.5.2.3	Minima Hopping *182*
	References *185*

7	**Perspectives and Outlook** *187*
7.1	Expansion through the Community *187*
7.2	Future Algorithm Developments *187*
7.3	Problems to Tackle – Discovery versus Design *188*

Index *191*

Preface

Knowledge of the atomic structure of a materials system is the key to understanding most of its properties, as well as the physical phenomena that can occur in that system. Even when the bulk crystal structure of a material is known or understood, that knowledge does not readily imply that we know the structure of atomic clusters or nanowires made of that particular material. We might venture a good guess for clusters with very large numbers of atoms, or in the case of very thick nanowires (whiskers), but most of the new and interesting phenomena will not happen in the regimes of large dimensions but at the nanoscale. At the nanoscale, often the structure and properties of materials have little to do with the structure of the bulk crystalline material! This realization has provided strong motivation for the development of methodologies aimed at finding the atomic configuration of nanostructures. If we analyze deeper the field of atomic structure determination, we would recognize, for example, that our old college textbooks provided only limited understanding of the correlation between the composition of an alloy or compound and its equilibrium structure: this situation was described a few decades ago as a "continuing scandal in physical sciences" (John Maddox, 1988). This challenge also sparked significant and long-standing efforts to predict the crystal structure from its composition. The two main impediments to crystal structure prediction and to the determination of the atomic configuration of nanostructures have been for a long time the lack of realistic models of atomic interactions and the lack of an efficient optimization scheme. Both of these impediments have by now been remedied for a respectable number of materials systems, and the time seems ripe to describe atomic structure prediction in a book. We are doing that here, focusing mainly on one method, the genetic algorithm. Genetic algorithms "mimic" the processes of the natural evolution to create progressively better-fit atomic structures, and eventually find or predict their equilibrium configurations. While one may argue that the natural evolution is overly simplified in this case, practice has shown that it is the simplification of dealing with real-space configuration (as opposed to their representation as genotype, akin to actual genes) that is responsible for obtaining solutions within a reasonable time and computational resources.

The main purpose of this book is to offer a beginning practitioner (often a graduate student) a detailed background in the way genetic algorithms for atomic structure prediction work. As noted in the book, we use the terms "genetic" and "evolutionary" interchangeably, although a few workers in the field tend to reserve the term "genetic" to describe the algorithm in the binary representation of the coordinate space. Confusions are unlikely, since nowadays the binary representation is hardly used anymore. When this book was commissioned by Wiley-VCH, the focus was only on nanostructures, clusters, and surfaces, hence the cover title. However, while this project progressed, so did the state of the field – culminating with the genetic algorithm solution of one of the most important problems in solid-state physics, that is, the prediction of crystal structure solely from the knowledge of its composition. We have included this important development in the book, so the reader obtains a timely and up-to-date view of the field.

While our aim is to provide a clear understanding of how the genetic algorithm works and of its strengths and successes, this book is by no means a review of all genetic algorithm works published so far. Given the rapid developments in the field, such a review would be both too long and incomplete. We have structured this book as a primer that intertwines basic general descriptions of the algorithms (e.g., the operations, the fitness functions, boundary conditions, where and how the algorithm can be applied, etc.) with full-detail applications to specific systems including crystals (3D), surfaces (2D), nanowires and nanotubes (1D), and certainly clusters (0D). We believe that the reader would benefit from this approach, as well as from the fact that all the examples or case studies have been scrutinized in the peer-review process (as technical journal publications) and contain virtually all working details of the approach and of the analysis. As such, we hope that the book provides significant understanding of the method so that any reader interested in applying the algorithm to his or her problem can do so with relative ease and confidence – whether or not his/her problem has already been covered in the book.

Throughout the years, we have been fortunate to work with a number of faculty colleagues, postdoctoral associates, and graduate students from Ames Laboratory, Iowa State University, and Colorado School of Mines, as well as from many other institutions. The work that is described in most detail (i.e., the full-detail applications) in this book has been performed alongside with them, and we gratefully recognize them here: R.M. Briggs, T.-L. Chan, F.-C. Chuang, T. Davies, D.M. Deaven, G.H. Gilmer, B. Harmon, M. Ji, B. Liu, N. Lu, D.E. Lytle, D.P. Mehta, J.R. Morris, M.C. Nguyen, C. Predescu, J.-L. Rodriguez-Lopez, V.B. Shenoy, K. Umemoto, R.M. Wentzcovitch, S. Wu, J. Zhang, and X. Zhao.

We thank them for their work and insights, and we look forward to our continued collaborations. Last but not least, we thank very much the editor of Physical Sciences Books at Wiley-VCH, Nina Stadthaus, who has guided this project with unmatched professionalism, patience, and dedication.

Golden, CO *Cristian V. Ciobanu*
Ames, IA *Cai-Zhuang Wang*
Ames, IA *Kai-Ming Ho*
September 2012

1
The Challenge of Predicting Atomic Structure

The atomic structure is the most important piece of information that is necessary when studying the properties of crystals, surfaces, interfaces, or nanostructures. If the bulk crystal structure of a material or compound is known, this may, under certain conditions, help in determining the structures of surfaces, clusters, or nanowires; however, such knowledge does not at all imply that we automatically know the structure of a surface or of a nanoparticle made out of that material. Often the surfaces and nanostructures adopt very intriguing atomic configurations, especially when their size is small (e.g., small number of atoms for clusters, thin diameter for nanowires, etc.). By now, the structure of many crystals is already known and usually taken for granted, sometimes to the point of considering a surface or a nanostructure as a simple truncation of the bulk material. As we see in other chapters, this is rarely the case at the nanoscale: the structure of atomic small clusters has little (usually nothing) to do with that of the bulk crystalline material!

Coming back to the issue of bulk crystal structure, on fundamental grounds, we have to recognize that retrieving the crystal structure from a given material or compound solely from knowing its composition is not an easy or a straightforward task: what determines, say, molybdenum to "choose" a body-centered cubic structure at normal conditions of temperature and pressure, when palladium has a face-centered cubic (fcc) structure and ruthenium adopts a hexagonal close-packed one? Why does NaCl, the common salt, adopt a structure in which both the fcc sublattices of Na and those of Cl are displaced along the side of the conventional cube, but CsCl adopts a different structure even though Cs is in the same group as Na in the periodic table? Granted, one can easily give a partial answer to this question on grounds that the Cs atom, although of same valence as Na, has a larger ionic radius and therefore would tend to have more Cl atoms around it than Na has. Still, what made NaCl adopt its specific structure (Figure 1.1) in the first place? Why are the sodium atoms in NaCl crystal arranged in an fcc structure, could they not have chosen a different arrangement? Some decades ago, John Maddox [1] phrased the problem of determining the crystal

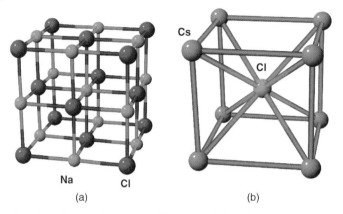

Figure 1.1 The crystal structures of NaCl (a) and CsCl (b).

structure from its composition as provocation also meant as a statement of fact (or perhaps the other way around):

"One of the continuing scandals in the physical sciences is that it remains impossible to predict the structure of even the simplest crystalline solids from a knowledge of their composition."

The scientific community has responded very well to that provocation: now, two and a half decades later, the development of global search algorithms coupled with the increase in available computing power allows us to make the crystal structure predictions from first-principles calculations, starting only with the desired composition. In this chapter, we are going to follow the main milestones that have led to such progress, and then describe the scope and organization of this book.

1.1
Evolution: Reality and Algorithms

What eventually has facilitated the best answer to Maddox's challenge are evolutionary algorithms or genetic algorithms (GAs) for structure prediction. We use these terms interchangeably, although only part of the scientific community does so. Methods to simulate materials thermodynamics already existed (such as simulated annealing [2]): those methods are highly meaningful from a fundamental standpoint. Not only do they simulate the thermodynamics of the material, but, in principle, they may double as ground-state structure prediction methods if the system is run at high enough temperatures and then slowly cooled down. However versatile this approach may seem at the first sight, in its early days it has not always led to ground-state structures due to the tendency to get stuck in metastable local minima. As it turns out, having the materials system go over energetic barriers in the potential energy surface (PES) is not an easy task: in fact, it is extremely time consuming and inefficient. Even if sufficiently slow cooling rates may be achieved for certain systems and ground state is reached, those rates do not serve as universal knowledge to be transferred to other materials.

What has significantly changed this situation are methods that can cross energy barriers with some ease, which have been continuously developed starting in early 1990s. Among those, the GA approaches have been developed now to the point that they make reliable predictions of cluster or crystal structures. In a genetic algorithm, we usually start with a pool of structures (genetic pool) that may or may not have anything to do with the real material structure. Only the nature of the atoms that compose those structures should be chosen as desired.

GAs are simple and generic methods that evolve the structures in this pool according to a certain energetic criterion (or cost function). The evolution proceeds via a set of genetic operations: through these genetic operations, new structures are created from the old ones, their cost functions are evaluated, and the new structures are considered for inclusion in the pool of structures depending on the values of their cost functions. If the energy (cost function) of a new structure is sufficiently low, then that newly created structure will be included in the genetic pool at the expense of an older and less energetically favorable one. It is obvious that the GA procedure, which will be described in more detail throughout this book, is guaranteed to lead to improvements in the structure in the pool. At the very worst, any newly created structure has higher energy than the old ones at any given point, in which case the genetic pool is never updated. However, this never happens if the initial genetic pool is initialized with random structures. The genetic operations, after a sufficient number of applications, will generate structures that are better than the random ones simply by virtue of keeping the pool updated to include structures with low energies.

In a way, the analogy with reality is very clear. Traits (atomic configurations) that are beneficial to the species are propagated in time through the genetic operations, while those that are not lead to creation of specimens that are unfit to survive and eventually die (structures are pushed out of the genetic pool). Although the principle of GA is strikingly close to that encountered in the actual Darwinian evolution of species, we note that in GA for structure optimization the parent structures can pass on not only traits that they themselves have been born with but also traits that they have acquired during the evolution: this is called Lamarckian evolution. In fact, this Lamarckian evolution coupled with the relative simplicity of atomic and molecular systems compared to actual genomes leads to structures that evolved sufficiently fast. In the evolution of species, there is no actual end to the process; that is, there is a continuous adaptation to the environmental conditions of life. However, when using GAs for atomic structure problems, we use them as a way to find a solution, usually a ground-state structure for some given conditions; that is, we use the lowest-energy structure for a given composition and then halt the procedure if the best structure in the genetic pool stops improving after a prescribed number of generations.

1.2
Brief Historical Perspective

The use of GAs as a structure optimization method can be traced back to the work of Hartke [3], who used it for small silicon clusters; another early use of the method was for small molecular clusters [4]. It is interesting to note that in order to keep the algorithm close to the real-life evolution, the structures in those early works were encoded as strings of 0s and 1s, strings that are interpreted as the genomes of the individual structures. Then, new structures would be created from gene splicing in which genetic operators (crossover, mutation) acted on binary strings, bit by bit. As a matter of historical perspective, we have to note that the encoding in the work of Hartke was not done by independently discretizing the space; rather, there was a prescribed set of geometric parameters associated with the cluster that was encoded as a binary string. Although there have been other works benefiting from GA searches in this binary encoding (called genotype representation) [5–7], there are already two problems to be recognized. First, the encoding of space requires a careful definition of certain geometric parameters. Second, because in this encoding GA does not fine-tune on local or global minima, in order to do that one would have to add a minimization step, in real space.

As a response to the first of these two problems, Zeiri [8] described the structures in a genetic pool not as binary strings, but as the collection of position vectors for the atoms. This description removed the artifacts associated with encoding and decoding, which stem from user-based decisions of the parameters to be encoded. This obviation of the requirement for encoding was an important conceptual step in the development of GA for cluster structure determination. Also in 1995, the next significant steps in the development of the algorithm came from Deaven and Ho [9], who (i) replaced the gene-splicing operations by real-space crossover operations in which spatial domains of two parent structures are combined together to form a new structure and (ii) introduced a local optimization (relaxation) step after each new structure was generated. The introduction of a local relaxation with every single structure considered for inclusion into the genetic pool changes the PES of the system, in a way that makes the algorithm deal only with local minima on the PES and nothing else. This is a significant simplification of the structure determination problem and is the key to the success of GA. There are still energy barriers to be crossed from between various local minima, but the cut-and-paste crossover operation of Deaven and Ho [9] is not hampered by these barriers. This is because the local minimization of coordinates can often alleviate most of the stresses created at the boundary between the two parental domains joined together; some new characteristics are acquired after the relaxation step, which is the reason why in Deaven–Ho development the GA is more akin to the Lamarckian evolution than to the Darwinian one.

After a few initial tests on carbon clusters [9], point charges [10], and Lennard-Jones systems [11], the GA in real-space representation with the cut-and-paste crossover operations has became mature and versatile enough to tackle systems of tens to hundreds of atoms. As such, the use of GA has spread with its application to finding the structure of clusters made of very different materials, such as silicon [12], germanium [13], metals (Al, Au, Ag, Ni, Zn, and others [14–20]), metallic alloys (Ni–Al, Cu–Au, and others [21–25]), oxides (MgO, TiO_2, SiO_2, and others [26–28]), water molecules [29,30], and many others. During this time of rapidly expanding applications of GA, it has become apparent that the method is very efficient and powerful. It has also become clear that, short of the dimensionality curse (by which any global structural optimization method fails if the number of atoms is increased sufficiently), the quality of the results or lack thereof is not affected by the GA procedure itself but, rather, by the interatomic potential model used in the determination of energies or cost functions. The results of GA searches based on empirical potentials are checked, when possible, at the end of the GA procedures against density functional theory (DFT) or tight-binding calculations. Often these checks reveal a significantly different rank ordering of the structures at the DFT level as compared to the empirical potential level. For this reason, GA global optimum searches are more and more often coupled with accurate tight-binding empirical potentials or with DFT calculations (see, for example, Ref. [12]).

With the standing of real-space GA well established as a method of efficiently locating global minima, the community has changed focus to systems other than clusters. In 2004, Ciobanu and Predescu cast the reconstruction of silicon surfaces as a problem of global optimization of the positions of the atoms at the surface and designed a parallel tempering annealing method to find the global minimum for high-index silicon surfaces [31]. Clearly, if a Monte Carlo-based global optimizer was successful, a GA one should also be able to find the optimal reconstruction of surface. And indeed, shortly thereafter a GA for finding surface reconstructions has been designed and results have been published for a variety of surface orientations of silicon [32–36]. The main idea with the GA for surfaces was the realization that the set of atoms subjected to GA operations can and should be different from the total number of atoms present in the systems: this is so because the structures are selected based on their surface energy, which requires many surface layers to relax in order to be correctly calculated. This way, the optimization is carried out over much fewer atoms than those needed to compute surface energies; in passing, we note that the same requirement holds true for the case of interfaces, for which case GA can also identify optimal structures [37,38]. Furthermore, conceptually, there is another key aspect of the algorithm development that was not present in the case of clusters: the number of atoms subjected to GA operations could be variable and does not have to be forced to remain constant [32]. This is because the surface energy defined per unit area and thus the number of atoms subjected to GA is not a predefined physical characteristic of the system (the way it is for clusters).

In parallel with the developments for surface reconstructions, the GA was applied to finding the structure of ultrathin nanowires [39–42]. In the case of metals, GA had minimal modifications with respect to its implementation for atomic clusters. Those modifications consisted, evidently, in the use of periodic boundary conditions along one spatial direction [39,40]. In the case of silicon nanowires (SiNWs), remarkable progress has been achieved in terms of the preparation and characterization of SiNWs, but atomic-level knowledge of the structure was still required for a complete understanding of the device properties of these wires. GA was envisioned as a tool to gain insight into the structure, but the role of passivation and the parameters associated with it were not readily clear and transferable potential models had yet to be tested against DFT data. Eventually, around 2006, these difficulties were shown to be manageable and GA has registered success in predicting the structure of passivated Si nanowires [41,42].

In 2006, Abraham and Probert refreshed the interest of the scientific community in using global optimizers for predicting crystal structures with a very creative article [43]. These authors showed that the cut-and-paste crossover of Deaven and Ho [9] can be used in the context of periodic boundary conditions in the three spatial dimensions provided it is executed on the fractional coordinates of the atoms and not on the Cartesian ones. Combined with the variable number feature [32], the real-space GA for crystal structure prediction of Abraham and Probert was successful in finding the global minimum energy configurations as well as in finding polymorphs. The algorithm required no prior assumptions regarding the size, shape, or symmetry of the simulation cell, and also no knowledge about the atomic configuration in the cell. Results on large Lennard-Jones systems with fcc- and hcp-commensurate cells showed robust convergence to the bulk structure from a random initial assignment and an ability to successfully discriminate between competing low-energy configurations. Abraham and Probert also coupled their GA with *ab initio* calculations and have shown the spontaneous emergence of both lonsdaleite and graphite-like structures in their GA search [43]. This, in effect, was a very clear answer to Maddox's challenge [1].

In June 2006, Oganov and Glass [44] also published a very similar GA development for crystal structure predictions, followed by Trimarchi and Zunger's development in 2007 [45]. The main advancements over Abraham and Probert are in regard to the efficiency of the algorithm. Since Oganov and Glass [44] couple their GA mostly to DFT calculations (as do Trimarchi and Zunger [45]), it becomes critical to have very clear cases in which new structures can be discarded without having a DFT relaxation performed on them; another factor that beneficially affects the time taken to success is seeding the initial pool of structures with known polymorphs. The work of Oganov and Glass culminated with the development of a USPEX (Universal Structure Predictor: Evolutionary Xtallography) code, which at present is available to users worldwide [46]. This code was interfaced with widely popular *ab initio* packages such as VASP [47] and SIESTA [48], and with GULP for atomistic calculations [49]. A plethora of applications of the GA for crystal structure prediction followed [50–65], effectively articulating a welcome and clear answer to Maddox's challenge.

1.3
Scope and Organization of This Book

The purpose of this book is to provide a beginner, most usually a graduate student, a solid background in the genetic algorithms for solving very diverse problems of atomic structure prediction. We use the terms "genetic" and "evolutionary" interchangeably, but the reader will notice the former term more often. We believe that this should not really create confusion, since nowadays the genotype representation of the method (i.e., the one in which atomic structures have to be binary encoded) does not offer solutions anywhere close to those obtained using the real-space representation in terms of robustness, efficiency, or achievable size of the problem with given resources. Other than mentioning it in historical context (such as that of the previous section) and in brief comparisons, the genotype representation is virtually absent from this book.

We describe in detail the inner workings of the real-space GA method in Chapters 2–5 for various materials systems and structural problems, guided by the developments described above. Specifically, we discuss the general method (Chapter 2), while considering all possible system dimensionalities from 0D to 3D. The specific applications to crystals and clusters are described in Chapters 3 and 4, respectively. Chapter 5, the most extended one in the book, contains applications to surfaces, interfaces, nanowires, and nanotubes. A short account of a few other popular global structural optimization methods is given in Chapter 6. Even though there are codes available [46], those do not necessarily tackle all problems regarding structure optimization; many others remain to be solved, as described in Chapter 7.

While we do aim to provide a clear understanding of the inner workings and of the strengths and successes of GA in real-space representation, this book is not a review of all GA works published so far. Rather, it is meant as a primer that also contains partially [9, 12, 13, 31] or fully described examples [10,32–35,37,41,63–69] from the work done over the years by our own groups at Ames Laboratory/Iowa State University and Colorado School of Mines or done through collaborations with colleagues from other institutions. The references in every chapter are selected judiciously so as to provide the reader a good picture of the current state of the field, but the list is certainly not exhaustive – the field grows too rapidly to even make attempts at completeness of citation lists. The book benefits from the fact that all the examples or case studies have been scrutinized in the peer-review process and contain virtually all working details of the approach and analysis – which should provide insights into the method to any reader interested in designing her or his own GA methodology for problems that may or may not be similar to the examples given.

References

1 Maddox, J. (1988) *Nature*, **335**, 201.
2 Kirkpatrick, S. (1983) *Science*, **220**, 671.
3 Hartke, B. (1993) *J. Phys. Chem.*, **97**, 9973.
4 Xiao, Y. and Williams, D.E. (1993) *Chem. Phys. Lett.*, **215**, 17.
5 Hartke, B. (1996) *Chem. Phys. Lett.*, **258**, 144.
6 Hartke, B. (2003) *Phys. Chem. Chem. Phys.*, **5**, 275.
7 Hartke, B., Flad, H.-J., and Dolg, M. (2001) *Phys. Chem. Chem. Phys.*, **3**, 5121.
8 Zeiri, Y. (1995) *Phys. Rev. E*, **51**, 2769.
9 Deaven, D.M. and Ho, K.M. (1995) *Phys. Rev. Lett.*, **75**, 288.
10 Morris, J.R., Deaven, D.M., and Ho, K.M. (1996) *Phys. Rev. B*, **53**, R1740.
11 Deaven, D.M., Tit, N., Morris, J.R., and Ho, K.M. (1996) *Chem. Phys. Lett.*, **256**, 195.

12 Ho, K.M., Shvartsburg, A.A., Pan, B., Lu, Z.Y., Wang, C.Z., Wacker, J.G., Fye, J.L., and Jarrold, M.F. (1998) *Nature*, **392**, 582.
13 Qin, W., Lu, W.C., Zhao, L.Z., Zang, Q.J., Chen, G.J., Wang, C.Z., and Ho, K.M. (2009) *J. Chem. Phys.*, **131**, 124507.
14 Lloyd, L.D., Johnston, R.L., Roberts, C. et al. (2002) *ChemPhysChem*, **3**, 408.
15 Chuang, F.C., Wang, C.Z., and Ho, K.M. (2006) *Phys. Rev. B*, **73**, 125431.
16 Michaelian, K., Rendon, N., and Garzon, I.L. (1999) *Phys. Rev. B*, **60**, 2000.
17 Zhang, W.X., Liu, L., and Li, Y.F. (1999) *Acta Phys. Sin.*, **48**, 642.
18 Wang, J.L., Wang, G.H., and Zhao, J.J. (2003) *Phys. Rev. A*, **68**, 013201.
19 Sun, H.Q., Luo, Y.H., Zhao, J.J. et al. (1999) *Phys. Status Solidi B*, **215**, 1127.
20 Wang, B.L., Zhao, J.J., Chen, X.S. et al. (2005) *Phys. Rev. A*, **71**, 033201.
21 Hsu, P.J. and Lai, S.K. (2006) *J. Chem. Phys.*, **124**, 044711.
22 Wang, J.L., Wang, G.H., Chen, X.S. et al. (2002) *Phys. Rev. B*, **66**, 014419.
23 Bailey, M.S., Wilson, N.T., Roberts, C. et al. (2003) *Eur. Phys. J. D*, **25**, 41.
24 Diaz-Ortiz, A., Aguilera-Granja, F., Michaelian, K. et al. (2005) *Physica B*, **370**, 200.
25 Darby, S., Mortimer-Jones, T.V., Johnston, R.L. et al. (2002) *J. Chem. Phys.*, **116**, 1536.
26 Hamad, S., Catlow, C.R.A., Woodley, S.M. et al. (2005) *J. Phys. Chem. B*, **109**, 15741.
27 Wang, C., Liu, L., and Li, Y.F. (1999) *Acta Phys.-Chim. Sin.*, **15**, 143–149.
28 Roberts, C. and Johnston, R.L. (2001) *Phys. Chem. Chem. Phys.*, **3**, 5024.
29 Hartke, B., Schutz, M., and Werner, H.J. (1998) *Chem. Phys.*, **239**, 561.
30 Guimaraes, F.F., Belchior, J.C., Johnston, R.L. et al. (2002) *J. Chem. Phys.*, **116**, 8327.
31 Ciobanu, C.V. and Predescu, C. (2004) *Phys. Rev. B*, **70**, 085321.
32 Chuang, F.C., Ciobanu, C.V., Shenoy, V.B., Wang, C.Z., and Ho, K.M. (2004) *Surf. Sci.*, **573**, L375.
33 Chuang, F.C., Ciobanu, C.V., Predescu, C., Wang, C.Z., and Ho, K.M. (2005) *Surf. Sci.*, **578**, 183.
34 Chuang, F.C., Ciobanu, C.V., Wang, C.Z., and Ho, K.M. (2005) *Jpn. J. Appl. Phys.*, **98**, 073507.
35 Ciobanu, C.V., Chuang, F.C., and Lytle, D.E. (2007) *Appl. Phys. Lett.*, **91**, 171909.
36 Ciobanu, C.V., Jariwala, B.N., Davies, T.E.B., and Agarwal, S. (2009) *Comput. Mater. Sci.*, **45**, 150.
37 Zhang, J., Wang, C.Z., and Ho, K.M. (2009) *Phys. Rev. B*, **80**, 174102.
38 Chua, A.L.S., Benedeck, N.A., Chen, L., Finnis, M.W., and Sutton, A.P. (2010) *Nat. Mater.*, **9**, 418.
39 Wang, B., Yin, S., Wang, G.H., Buldum, A., and Zhao, J.J. (2001) *Phys. Rev. Lett.*, **86**, 2046.
40 Wang, B.L., Wang, G.H., and Zhao, J.J. (2002) *Phys. Rev. B*, **65**, 235406.
41 Chan, T.L., Ciobanu, C.V., Chuang, F.C., Lu, N., Wang, C.Z., and Ho, K.M. (2006) *Nano Lett.*, **6**, 277.
42 Lu, N., Ciobanu, C.V., Chan, T.L., Chuang, F.C., Wang, C.Z., and Ho, K.M. (2007) *J. Phys. Chem. C*, **111**, 7933.
43 Abraham, N.L. and Probert, M.I.J. (2006) *Phys. Rev. B*, **73**, 224104.
44 Oganov, A.R. and Glass, C.W. (2006) *J. Chem. Phys.*, **124**, 244704.
45 Trimarchi, G. and Zunger, A. (2007) *Phys. Rev. B*, **75**, 104113.
46 http://han.ess.sunysb.edu/~USPEX/.
47 Kresse, G. and Furthmüller, J. (1996) *Phys. Rev. B*, **54**, 11169.
48 Soler, J.M., Artacho, E., Gale, J.D., Garcia, A., Junquera, J., Ordejon, P., and Sanchez-Portal, D. (2002) *J. Phys.: Condens. Matter*, **14**, 2745.
49 Gale, J.D. (2005) *Z. Kristallogr.*, **220**, 552.
50 Oganov, A.R., Glass, C.W., and Ono, S. (2006) *Earth Planet. Sci. Lett.*, **241**, 95.
51 Oganov, A.R., Ono, S., Ma, Y.M., Glass, C.W., and Garcia, A. (2008) *Earth Planet. Sci. Lett.*, **273**, 38.
52 Gao, G.Y., Oganov, A.R., Bergara, A., Martinez-Canales, M., Cui, T., Iitaka, T., Ma, Y.M., and Zou, G.T. (2008) *Phys. Rev. Lett.*, **101**, 107002.
53 Oganov, A.R., Chen, J.H., Gatti, C., Ma, Y.Z., Ma, Y.M., Glass, C.W., Liu, Z.X., Yu, T., Kurakevych, O.O., and Solozhenko, V.L. (2009) *Nature*, **457**, 863.
54 Ma, Y.M., Oganov, A.R., Li, Z.W., Xie, Y., and Kotakoski, J. (2009) *Phys. Rev. Lett.*, **102**, 065501.
55 Li, Q., Ma, Y.M., Oganov, A.R., Wang, H.B., Wang, H., Xu, Y., Cui, T., Mao, H.K., and

Zou, G.T. (2009) *Phys. Rev. Lett.*, **102**, 175506.

56 Oganov, A.R., Ma, Y.M., Xu, Y., Errea, I., Bergara, A., and Lyakov, A.O. (2010) *Proc. Natl. Acad. Sci. USA*, **107**, 7646.

57 Xie, Y., Oganov, A.R., and Ma, Y.M. (2010) *Phys. Rev. Lett.*, **104**, 177005.

58 Zhou, X.F., Dong, X., Zhao, Z.S., Oganov, A.R., Tian, Y.J., and Wang, H.T. (2012) *Appl. Phys. Lett.*, **100**, 061905.

59 d'Avezac, M. and Zunger, A. (2007) *J. Phys.: Condens. Matter*, **19**, 402201.

60 Trimarchi, G. and Zunger, A. (2008) *J. Phys.: Condens. Matter*, **20**, 295212.

61 Trimarchi, G., Freeman, A.J., and Zunger, A. (2009) *Phys. Rev. B*, **80**, 092101.

62 d'Avezac, M., Luo, J.W., Chanier, T., and Zunger, A. (2012) *Phys. Rev. Lett.*, **108**, 027401.

63 Ji, M., Umemoto, K., Wang, C.Z., Ho, K.M., and Wentzkovitch, R.M. (2011) *Phys. Rev. B*, **84**, 220105(R).

64 Wu, S., Umemoto, K., Ji, M., Wang, C.Z., Ho, K.M., and Wentzkovitch, R.M. (2011) *Phys. Rev. B*, **83**, 184102.

65 Nguyen, M.C., Zhao, X., Ji, M., Wang, C.Z., Harmon, B., and Ho, K.M. (2012) *J. Appl. Phys.*, **111**, 07E338.

66 Chuang, F.-C., Liu, B., Wang, C.Z., Chan, T.L., and Ho, K.M. (2005) *Surf. Sci.*, **598**, L339.

67 Briggs, R.M. and Ciobanu, C.V. (2007) *Phys. Rev. B*, **75**, 195415.

68 Davies, T.E.B., Mehta, D.P., Rodriguez-Lopez, J.L., Gilmer, G.H., and Ciobanu, C.V. (2009) *Mater. Manuf. Process.*, **24**, 265.

69 Ji, M., Wang, C.-Z., and Ho, K.-M. (2010) *Phys. Chem. Chem. Phys.*, **12**, 11617.

2
The Genetic Algorithm in Real-Space Representation

This chapter, as the remainder of the book, focuses strictly on real-space representation solutions of atomic structure problems. As mentioned in Chapter 1, in this representation, the solution is sought directly in terms of spatial coordinates of the atoms in a given structure, without passing through a binary representation. We will address, with certain generality, the operational way in which atomic structure problems are solved using genetic algorithms (GAs). It is important to understand in detail the options that we have, or that we can produce, in terms of implementing the main GA operations, i.e., selection, crossover, and mutation. The reason for this is the practical observation that there is not, for example, a general best way of evolving the genetic pool or a general best way of performing a crossover, and so on. Many of the options or choices available do not influence the convergence tremendously if they are exercised wisely. There is certainly a general procedure for carrying out genetic algorithm optimization, but the components of it have to be modified according to the optimization problem; as an obvious example, one needs to change the quantity to optimize when dealing with surfaces, as opposed to clusters. Moreover, there are many sets of sensible options in setting up GA operations, and we describe in this chapter one of them while pointing out at times ways in which the practitioners can modify such set. Although the overall complexity of a problem is determined to a large extent by the number of degrees of freedom of the atomic system considered, structural problems can still vary widely in terms of their level of difficulty (assessed, for example, by the computational time taken to reach a solution using same computing resources); the level of difficulty can be affected by boundary conditions or by interactions prescribed for atoms. Interestingly, the level of difficulty can also depend directly on how "smart" the crossover operations are or on how the genetic pool is advanced from one generation to the next. This is to say, a given optimization problem can prove lengthy to solve if we have a "poor" crossover or a "poor" updating of the genetic pool. But how do we know, *a priori*, what a poor or smart crossover is for a given problem? Strictly speaking, we do not! However, there are certain guidelines for assessing the suitability or performance of a set of genetic operations. After a brief sampling of several important atomic structure problems (Section 2.1), this chapter addresses, in turn, the stages of genetic algorithms and the typical operations (Sections 2.2–2.6) in the context of various structural problems that currently arise in nanoscience, physical chemistry, or condensed matter physics.

2.1
Structure Determination Problems

Let us state precisely, for a few important cases, the atomic structure problems that should be tackled by global optimization methods, and in particular by genetic algorithms. Such statements, which are at the heart of this book, are important not merely for the sake of completeness, but for understanding the details of the evolutionary procedure that should be adopted toward their solutions.

2.1.1
Cluster Structure

For a number n of atoms of a given atomic species, determine the structure that this collection of atoms adopts.

The determination of the structure is based on changing (optimizing) the atomic coordinates so as the total formation energy per atom f adopts the lowest possible value. The formation energy f is defined as

$$f = \frac{\Delta E}{n} \equiv \frac{E - n\mu}{n}, \tag{2.1}$$

where ΔE is the difference between the total energy E of the bound system (cluster) and $n\mu$ is the product of the number of atoms and the chemical potential μ of an atom in the given reference state. Since optimizations are carried out at $0\,K$, E is actually the total potential energy of the system. The number of degrees of freedom to optimize should be $3n$, since there are three (Cartesian) atomic coordinates for each atom in the cluster; however, since clusters that differ only by a spatial rotation or translation should not be considered as having different structures, the number of degrees of freedom is actually $3n - 6$. Note that for the optimization of an n-atom cluster, it would be sufficient to compare the total energy E of various n-atom structures, as opposed to the formation energy per atom. Still, using the formation energy per atom f instead of the total energy E allows us to compare the stabilities of clusters with different numbers of atoms, and thus decide if there is a strongly preferred size. For example, the optimization of clusters of carbon atoms shows the size $n = 60$ significantly more stable than the clusters with numbers of atoms that differ only a little: the $n = 60$ cluster adopts the well-known buckyball structure in which hexagons (six-atom rings) and pentagons (five-atom rings) adjoin to form a near-spherical structure in which every carbon atom is threefold coordinated (Figure 2.1).

The buckyball, discovered experimentally by Curl, Kroto, and Smalley – a discovery for which they were awarded the 1996 Nobel Prize in Chemistry, is a particularly stable form of carbon that appears in hot carbon plasmas [2]. Carbon is significantly richer in terms of displaying polymorphic crystal structures than other group IV elements such as silicon or germanium. Optimization of carbon clusters with lower numbers of atoms ($n < 60$) shows that, in the order of increasing n, they tend to form single-ring structures and then multiring configuration whose shape

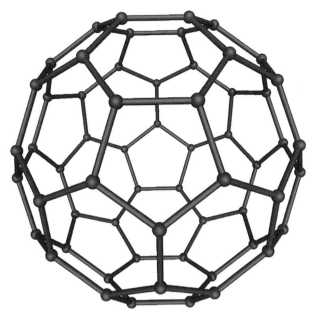

Figure 2.1 Buckyball, the ground-state structure of a 60-atom carbon cluster. This is the first report of successful real-space genetic algorithm optimization applied to clusters [1].

can be flat, bowl-like, and cap-like (Figure 2.2) [3]. These structures were obtained by local relaxations starting from different initial configurations, in order to access more than one possible isomer; energies were computed at the level of reactive Brenner potential [4]. Some of the reported ground-state structures shown in Figure 2.2 can certainly be considered under debate, as shown by comparisons with, for example, Refs [5–7]. As an immediate example of the debate, structures with $21 \leq n \leq 30$ take different configurations when the optimization method is aimed at the global minimum structure: using the same reactive potential [4], Wang et al. show that their quasi-dynamics-based global method leads to cage-like structures (Figure 2.3) for n values between 21 and 30.

Without aiming to solve or to exacerbate controversies in ground states of carbon clusters, we point out that there are two major reasons cluster structures (or, in fact, of any atomic structures) can come under debate. The first reason has to do with the type of interatomic potential used to calculate cluster energies: classical (empirical), tight-binding (TB), or pseoudopotentials developed for density functional theory (DFT) calculations, each having different versions with various levels of applicability or transferability. The second reason has to do with the optimization procedure. Even when same atomic interaction models are used, the ground-state and metastable structures can still differ depending on how robust the search for global minimum structure is. It is the case, for example, of the comparison mentioned above between the work of Kosimov et al. [3] and Wang et al. [7], that is, the comparison of Figures 2.2 and 2.3 for $21 \leq n \leq 30$.

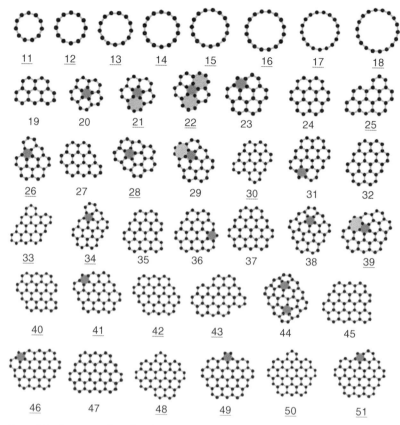

Figure 2.2 Structures of small carbon clusters for different values of n, obtained from local relaxations of different starting configurations at each value of n. (Adapted from Ref. [3], with permission from American Physical Society.)

Robust searches with well-transferable atomic interactions have also been published for silicon clusters [8] and germanium clusters [9]. Such searches reveal similarities and differences between structures of group IV elements with same numbers of atoms and offer insight into the growth patterns for these clusters [10].

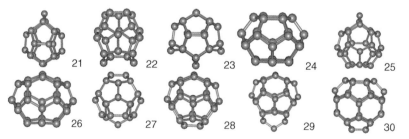

Figure 2.3 Structure of C_n clusters ($21 \leq n \leq 30$) obtained by a quasi-dynamics global search method. (From Ref. [7], with permission from American Physical Society.)

Initial theoretical predictions for clusters with $n > 8$ were diverse and contradictory, and none of them seemed particularly consistent with experimental data. However, clusters obtained via genetic algorithms coupled with DFT calculations of binding energy for the size range of $n = 12 - 18$ are consistent with experiments in terms of ionic mobilities determined for these global minima structures (Figure 2.4). As seen

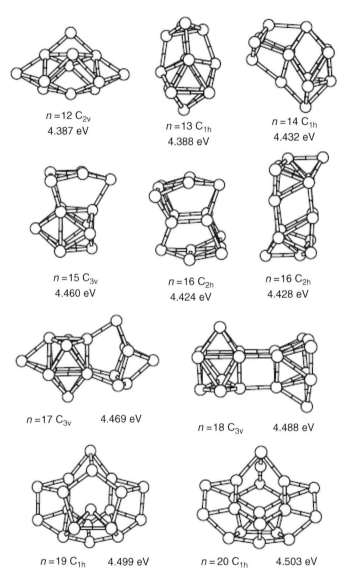

Figure 2.4 Global minima of Si_n ($12 \leq n \leq 20$) obtained via a genetic algorithm coupled with DFT calculations, with binding energy per atom indicated. The two geometries shown for Si_{16} are nearly degenerate. (From Ref. [8], with permission from Nature Publishing Group.)

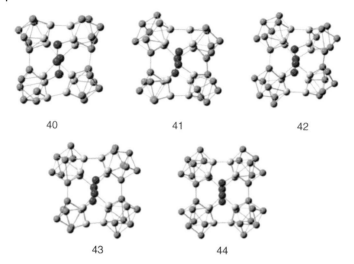

Figure 2.5 Plate-like Ge clusters obtained via a genetic algorithm combined with DFT calculations. Shades distinguish between the linker (core) and the corners of the plate-like structures. (Adapted from Ref. [9], with permission from American Institute of Physics.)

in Figure 2.4, these clusters are built upon a Si_9 trigonal prisms. At larger sizes, the global search predicts that near-spherical cage geometries become more stable than trigonal prisms.

In comparison, germanium clusters can form plate-like structures that consist in four Ge_9 or Ge_{10} subunits and a Ge_4 core that acts as a linker connecting the four subunits (refer to Figure 2.5). Such optimized geometry of Ge clusters already reveals significant differences in growth patterns between C, Si, and Ge, which could be exploited in practice since variations in sizes in the regime of 1–100 atoms correspond to significant differences in electronic and optic properties.

The field of cluster structure determination and size-dependent properties (including stability, optical, electronic, and mechanical properties) has grown tremendously from the point of view of both algorithm development and applications, and now covers a wide spectrum of materials, for example, metallic clusters, metallic alloy clusters, ceramic clusters, molecular clusters, and others, and we leave the more in-depth discussion of these types of clusters for Chapter 4.

2.1.2
Crystal Structure Prediction

Determine the structure of a crystalline material made of a known set of atomic species, with a given composition.

In this problem, it is not sufficient to optimize the atomic positions (up to an arbitrary translation), but we need to optimize the shape and size of the simulation box that model the entire crystal through the use of periodic boundary conditions.

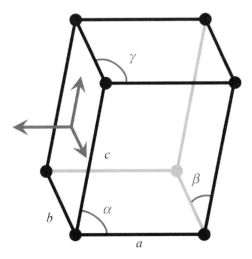

Figure 2.6 General shape of the periodic simulation cell, showing the dimensions (a, b, c) and angles (α, β, γ) that are to be optimized along with the positions of the atoms inside it (not shown). The arrows on one face of the simulation cell show components of the stress tensor that can be applied during the variable-cell relaxation; to avoid cluttering of the figure, these stress components are shown only on one face of the simulation cell.

Since the directional angles (α, β, γ) and the lengths of the simulation box (a, b, c) (see Figure 2.6) have to be optimized and since spatial rotations are not symmetry operations anymore, the number of degrees of freedom is not $3n - 6$ as in the case of clusters, but $3n + 3$. As we can expect, the optimization of the unit cell does not appear to be on the same footing as the optimization of atomic coordinates and requires a careful consideration of the crossover operations so as to allow for the mixing (crossover) of parent structures that have different shapes. Two parent structures that have the same simulation cell can be readily subjected to the crossover operations, which mix parts of the atomic structures of the two parents. However, two parent structures having simulation cells different in shape and size cannot be immediately subjected to crossover operations unless they are made compatible in some way. We will discuss this aspect of the optimization in Section 2.4. For now, it is sufficient to point out a tremendous advantage of having to optimize the supercell parameters (lengths and angles): this optimization allows us as users to impose a stress tensor with prescribed components σ_{ij} $(i, j = 1, 2, 3)$ on the supercell. Relaxation of the supercell under the condition of imposed stress tensor (which is usually some applied hydrostatic pressure) is a routine task in most current software packages (VASP [11], LAMMPS [12], etc.), so it does not require separate coding or implementation. The advantage is that this way we are able to predict high-pressure phases of materials, a computational feat that was achieved only recently for carbon phases in the range 0–2000 GPa (shown in Figure 2.7), sulfur at 12 GPa [13], silica in the TPa regime [14], and other materials.

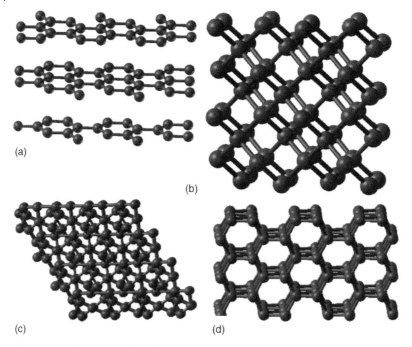

Figure 2.7 Carbon phases identified by Oganov and Glass using a genetic algorithm coupled with DFT calculations. (a) Graphite at 0 GPa. (b) Diamond, stable up to 1000 GPa. (c) bc8 structure stable above 1000 GPa. (d) Lonsdaleite, metastable at 100 GPa. (Adapted from Ref. [13], with permission from American Institute of Physics.)

In addition to the atomic coordinates and the parameters of the periodic supercell, we also do not know *a priori* the total number of atoms in the supercell. We may know the concentration of each component (according to the statement of the problem), but this does not mean that we know the total number of atoms. This is an important point because the physical quantity that describes the prediction of crystal structures should allow for comparison of crystalline structures with same composition but not necessarily with the same total number of atoms. For very low external pressures applied, we can still use the formation energy per atom f, only then we have a reference state for each particular component:

$$f = \frac{\Delta E}{n} = \frac{E - n_1 \mu_1 - n_2 \mu_2}{n}, \qquad (2.2)$$

where n_1 and $n_2 = n - n_1$ are numbers of atoms of type 1 and type 2, respectively, and μ_1, μ_2 are the chemical potentials corresponding to the reference states of pure components. Given the statement of the problem, n_1 is calculated according to the prescribed concentration. Certainly, Equation 2.2 can be generalized for more than two atomic components or simplified readily to one atomic component depending on the actual practical problem to solve. A robust solution of the problem should, ideally, not only predict the lowest f crystal structure but also the polymorph structures; the initial efforts toward polymorph prediction have been successful as we shall see in Chapter 3. Under conditions of applied pressure p, the quantity to be minimized (cost function) should not be the formation energy per atom (Equation 2.2) anymore, but rather the enthalpy of formation per atom for single species:

$$h = \frac{\Delta E + pV}{n} = \frac{E - n_1\mu_1 - n_2\mu_2 + pV}{n}, \tag{2.3}$$

where V is the volume of the unit cell. We can also certainly use the enthalpy of formation per unit formula of the compound (recall that the optimization occurs at prescribed composition): this quantity is readily computed since the applied pressure p is known and the volume V of the simulation cell can be readily computed after the variable cell relaxation.

2.1.3
Surface Reconstructions

For a given surface orientation of a known crystalline material, determine the positions of the atoms at the surface.

The positions of the atoms at the surface define the so-called surface reconstruction; these positions change from their bulk truncation values because the atoms at the surface can be reactive due to missing neighbors. This is a very fundamental problem in surface science, and rather a complex one at least in the case of semiconductor surfaces. For semiconductor surfaces, the locations of the atoms are determined not only by the need to minimize the number of dangling bonds at the surface but also by the requirement of minimizing the stress in the surface bonds built upon during the reduction of the number of dangling bonds. The energetic quantity that captures these two competing requirements is the surface energy per unit area, γ (or surface energy, for short):

$$\gamma = \frac{E - n_\mathrm{m}\mu}{A}, \tag{2.4}$$

where E is the total energy of the n_m atoms in the computational cell that are allowed to move, μ is the chemical potential of an atom in the (known) crystalline bulk state, and A is the surface area of the supercell. As shown in Figure 2.8, the atoms that are allowed to move, n_m, are not necessarily the same atoms whose positions are optimized via the genetic algorithm operations. This is because the reconstruction itself is contained in a couple of layers at the surface, but the relaxation due to surface bonding propagates deeper into the crystalline material.

This type of optimization problem brings to our attention the need to provide material with bulk character that is solely used for relaxation, without which the surface energy γ would be incorrectly determined. It emphasizes the idea of a template as a necessary construct for evaluating energetic quantities required to assess the fitness of a candidate structure. Templating is necessarily absent in the case of clusters and crystals, but should be considered carefully for surfaces and interfaces, as well as for problems dealing with nanoscale inclusions in a matrix material.

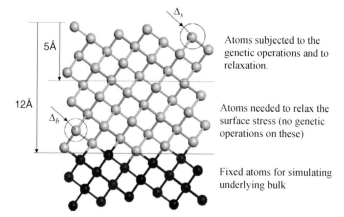

Figure 2.8 Surface slab with different operational zones marked. Upper region, typically 5 Å thick, is subjected to genetic algorithm operations; middle region is used only for relaxations; and bottom atoms are fixed and simulate the underlying bulk environment. The n_m atoms allowed to move, that is, entering in Equation 2.4, are those in the upper two regions combined. (Adapted from Ref. [16], with permission from American Physical Society.)

The first implementation of real-space GA to surface reconstruction problems [15] revealed that even with the use of an empirical potential for silicon [17], the stablest reconstruction of Si(105) surface is found using rather short GA runs. The structure, shown in Figure 2.9 along with the lowest surface energy across the population of structure as a function of "time," is direct evidence that genetic algorithms drastically reduce the chance of missing the correct physical surface reconstruction for high-index semiconductors surfaces, which imminently happens when heuristic approaches are used (see, for example, Ref. [18]). More

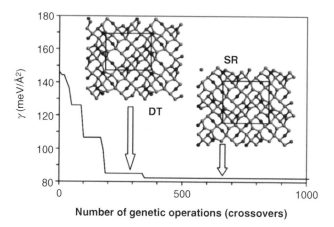

Figure 2.9 Lowest surface energy of a Si(105) slab during a GA optimization coupled with empirical potential. The lowest-energy structure is marked SR. (Adapted from Ref. [15], with permission from Elsevier.)

examples on GA for surfaces and other templated structure problems will be discussed in Chapter 5.

2.1.4
Range of Applications

At least in terms of the numbers of scientific works investigated and published so far, the two most important atomic structure optimization problems are the optimization of atomic clusters and the crystal structure prediction starting from a given composition of a material. While in the previous sections we described three major structural problems (optimization of clusters, crystal structures, and surface reconstructions), we have so far left aside the less studied issue of 1D nano-structures. Are there other important or potentially important problems that we have omitted?

In this section we try to give an exhaustive enumeration of the atomic optimization problems that can be solved via real-space genetic algorithms. In order to create this exhaustive description, we need to consider the following issues (ingredients):

a) Constituent material (single-element semiconductor, metal, others; molecular, alloy, compound, etc.)
b) Dimensionality of the system, as given by the number of directions with periodic boundary conditions that are involved in some way in the optimization procedure.
c) Total number of atoms, and the number of atoms of each species.
d) Surface passivation.

Figure 2.10 summarizes these possible structural problems that emerge as combinations between issues (a)–(d) listed above. Most of these problems are well defined, with very few obvious exceptions (such as a 3D crystal structure problem cannot have any passivating species because the bulk crystal exposes no surface to be passivated). As mentioned, not all problems are equally important, but most can become instrumental as part of other problems that require knowledge of

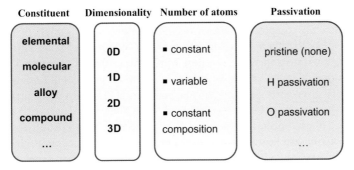

Figure 2.10 Range of atomic structure optimization problems stemming from different combinations of dimensionality, total number of atoms, composition, and passivation.

atomic structure. For example, knowledge of the structure of 1D atomic steps is necessary for understanding surface morphology, diffusion, or growth; the structure of a nanowire is important for determining its optoelectronic characteristics; the structure of a metallic cluster templated in a 3D matrix or on a 2D crystal surface is important for determining its catalytic properties with respect to certain desired reactions; and so on.

The issue of dimensionality may require certain clarification. For example, the structure of clusters (Section 2.1.1) does not need any periodic boundary conditions because the cluster cannot interact with its periodic images. The cluster problem is always 0D, with the "definition" given above; even if periodic boundary conditions are imposed (only because most simulation codes already operate with periodic boundary conditions), that would be done only after sufficient vacuum spacing has been inserted in all dimensions. The crystal structure prediction is always 3D, and most time one would need to optimize simultaneously the parameters of the computational cell (three angles and three periodic lengths) along with the atomic coordinates.

The number of atoms, listed as issue (b) above, is not just another variable but has to be handled differently depending on the problem given. In the case of atomic clusters, it is important that this number is kept constant: for single-species cluster, this number of atoms is a key characteristic of the cluster known prior to the optimization (the type of atom is the other important characteristic). Allowing it to vary freely during the crossover operations would yield nothing but continuously increasing clusters with structures tending toward the known bulk crystal structure and with properties that move away from those of small or medium clusters that are under investigation. If a size-dependent study is desired, then the structure, stability, and properties of each size n should be studied separately and the results should be compared at the end. At times, however, it is very advantageous to let the number of atoms vary: it is the case of surface reconstructions (2D) and of crystal structure prediction (3D), in which cases it was shown that convergence is made possible or at least improved to some degree [15,19]. Usually, composition (in the case of crystal structure of compounds or alloys) should be kept constant during optimization, otherwise the results do not quite correspond to a well-defined problem. However, as we shall see in Chapter 3, it is possible to determine all stable compositions of an alloy and their corresponding structures: if this is the goal, then it can only be achieved if we allow the composition to vary during the run.

Finally, we point out that there are structural problems (covered by Figure 2.10, but not made obvious in it) in which we deal with optimization of "templated" structure. For example (Figure 2.8), as we have seen in the 2D problem of finding reconstructions of a given surface orientation, the positions of the atoms in the first few layers at the surface are to be determined while they maintain contact with a fixed substrate slab (the template). The periodic dimensions are characteristic to the surface orientation and are kept fixed, so as to impart surface orientation to the atoms that are subjected to optimization via genetic algorithms. Another template problem that we consider 0D is the optimization of a cluster of given atomic species and size in a matrix of a different material: although in this case the matrix will require three periodic boundary conditions, the optimization of the cluster inside of it will never interfere with these conditions, which is why we consider this "templated" problem a 0D one.

2.2
General Procedure

The general procedure was described in Chapter 1, but it is instructive to give a graphical representation of it before going into details of each stage and each operation. This graphical representation is shown in Figure 2.11. The procedure starts from a "generation zero" genetic pool in which a number N_p of structures (population) are made of either randomly positioned atoms or from the output of a previous run, or by employing some other user choice. The reason for having several

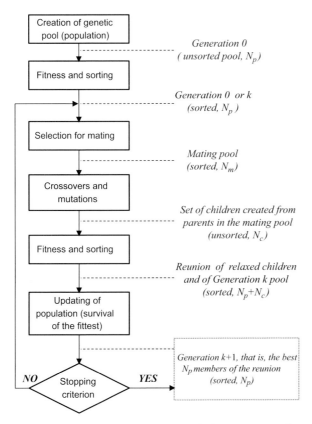

Figure 2.11 General flowchart illustrating the basic processes of a genetic algorithm. The changes brought to the genetic pool (population) after each process are described on the right-hand side.

choices for the generation zero is to check that the final output is not sensitive to the starting configurations.

Following the flowchart in Figure 2.11, the structures in the starting pool are relaxed via a local optimization procedure (steepest descents or conjugate gradient) and a cost function is assigned. This is the process called "fitness and sorting" in the diagram Figure 2.11: fitness means relaxation to the local energy minimum and assignation of an energy-related cost function or of a fitness function (we will clarify shortly the relation between these functions), and sorting means rank ordering the members of the populations (structures) from the best (less costly, most fit) to the worst (most costly, least fit). In principle, we could assign a fitness without relaxation to a local minimum, but such relaxation is necessary because it reduces drastically the coordinate space that is searched using the GA: through relaxation, only local minima on the potential energy surface (PES) of the system are being considered, as opposed to all possible configurations on the PES. The cost function (as opposed to fitness function) is to be minimized during the algorithm, and can be the formation energy per atom f defined for problems such as those described in Sections 2.1.1 and 2.1.2, or surface energy γ for surface reconstruction problems (Section 2.1.3), or other energetic quantity that is physically motivated by the problem at hand. For a global minimum (as opposed to global maximum) search, it appears more natural to deal with minimizing an energetic quantity (cost function, such as f or γ) than to deal with maximizing a fitness function. However, this is purely a matter of choice because the fitness function in this case could simply be the negative of the cost function; for example, we may define as a fitness function the binding energy per atom (in the case of clusters and crystals), which would simply be the negative of the formation energy per atom f defined in previous sections. Other fitness functions can be defined as well, with the purpose of spreading out their range and distinguishing more clearly between the structures in the pool. After relaxations into local minima, the structures in the pool are rank ordered from the lowest cost function (most fit) to the highest (least fit) and the algorithm proceeds with the selection of parents and genetic operations.

2.3
Selection of Parent Structures

Based on their fitness function, parents are selected to produce an offspring or child structure. This selection of the parents for reproduction (also known as crossover or mating) can be done in many ways. The most simple-minded selection of the parents is the *random selection*, in which any two structures in the genetic pool are equally likely (regardless of their fitness) to be selected for crossover to produce new offspring. This is not particularly efficient because "bad" subsets of clusters derived from least fit parents can survive longer periods during the evolution of the algorithm. Since this selection is random, eventually the good, fit parents will be selected for mating, so they will end up contributing good structural motifs (genes) to the population during the optimization procedure.

Selection leads to the creation of a mating pool, which is a subset of the initial genetic pool containing members that are used for creating new structures (offspring). Since this mating pool is the basis for new generations of structures, it is desirable to have better, more fit, configurations in the mating pool rather than random ones. Therefore, a better way to perform the selection of the parent structure in the mating pool would be one that explicitly takes into account the cost or fitness functions. For example, in their seminal paper, Deaven and Ho [1] choose the probability p_i of an individual structure i to be selected for mating as

$$p_i \propto exp[-f_i/T_c], \tag{2.5}$$

where f_i is the formation energy of structure i and T_c is the crossover "temperature," an energetic quantity roughly set to the range of energies f_i in the genetic pool. To be precise, they used the total energy per atom, but that is inconsequential for our discussion of Equation 2.5. This equation, which is chosen so as to closely resemble the Boltzmann distribution of a system on various single-particle energy levels f_i, introduces the idea of defining a fitness function other than the negative of the formation energy. To this end, and for the sake of having simulations closer in spirit to evolutionary procedures, we can require that the probability of selecting a parent for mating be proportional to its fitness F_i – fitness that we would define simply as [20]

$$F_i = \exp[-f_i/T_c]. \tag{2.6}$$

Thus, Equation 2.5 can be recast as

$$p_i = \frac{F_i}{\sum_{j=1}^{N_p} F_j}, \tag{2.7}$$

so the selection probabilities p_i are explicitly determined by the fitness values F_i. The selection probability p_i proportional to fitness F_i described here is often called *roulette-wheel selection* for the following reason. If we arrange all fitnesses in a pie chart (the roulette) in such a way that the angle at the center is proportional to the fitness F_i, then the probability of a marble ball to fall on a given sector is proportional to the angle of that sector – that is, is given by Equation 2.7. It may be interesting to note that the reader might see in the literature a difference between Boltzmann selection and roulette-wheel selection: such difference (which we have certainly not made in this section) only appears if the fitness function F_i is defined in a manner different from Equation 2.6.

Another possible way to select structures for the mating pool is the *tournament selection*. In this type of selection, every parent is selected as a result of a tournament between a prescribed number k of members of the genetic pool: the winner of the tournament is the structure with the highest fitness value among the k randomly chosen structures. We can define, loosely, the selection pressure as being the degree to which better fit individual structures are selected for the mating pool. For $k = 1$, the tournament selection is nothing else than the random selection, and the selection pressure is zero. For very large k, we end up in the limit of roulette-wheel selection, where selection probability is proportional to fitness. In principle, it is very important to be able to vary the selection pressure, because at low k values the algorithm may take a very long time to converge, whereas at high k values the algorithm converges too fast and it may do so to a suboptimal solution. The key is to allow for sufficient diversity, and often intermediate values of k can do just that. In practice, if we carry out the GA using the roulette-wheel selection and the algorithm has converged (i.e., the most fit structure stops changing), then we can loosen the selection pressure somewhat (i.e., decrease k) and continue the run in order to check if indeed convergence has been reached. Given its simplicity of concept and ease of implementation, tournament selection can and should always be used to cover the possibilities between the random and roulette-wheel selections.

2.4
Crossover Operations

The crossover operations (often also called mating operations or heredity operations) determine how new configurations (children, offspring) are created from two selected parents. In the binary representation discussed briefly in Chapter 1, two parent structures are represented by binary strings (chromosomes) and can be combined in rather simple ways to create a new binary string (Figure 2.12). One such way is to define a cutting point, before which the bit values are taken from one parent and after which the values can be taken from the other parent chromosome. Similarly, as seen in Figure 2.12, we can define a two-point crossover operation. One can also define a uniform crossover (not shown in Figure 2.12), in which one can prescribe a certain probability for each bit to be inherited from one parent or from the other. Each type of crossover operation (either in the binary representation or in real space) is able to create more than one child, but in the context of atomic structure problems we usually only deal with one child per crossover; in other words we strictly define which side of the cut(s) comes from the

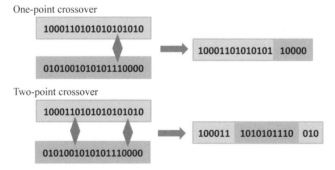

Figure 2.12 Possible crossover operations in the binary representation: each parent is represented as a binary string, and children can be created using one-point or two-point crossovers; the crossover points are indicated by diamonds.

2.4.1
Cut-and-Splice Crossover in Real Space

The binary representation has turned out to be inefficient for optimization of atomic clusters or crystal structures. In the spatial representation, the original crossover operation was very similar – often called cut-and-paste or cut-and-splice. In the real-space crossover [1], two parent structures are cut by the same plane and a child structure is created by combining the part of the first parent located on one side of the cutting plane with the part of the second parent situated on the other side of the plane (Figure 2.13). For a consistent operation that produces contiguous systems, the parent structures have to be translated with the origin of the coordinate system in their center, but, in fact, the cutting plane does not have to pass through the geometrical centers of the clusters. Allowing the plane to be situated at a random distance d from the center of the parent clusters (with d values between zero and the approximate radius of the cluster) can enable the algorithm search wider regions of the configuration space. Other possibilities, borrowed from the binary representations of the crossovers, would be the use of more than one plane to cut cluster structures or to use nonplanar cuts.

As we can see in Figure 2.13, the crossover operation is a construct without a physical basis: there is no single reaction or process that could combine two random unoptimal C_{60} clusters into a buckyball. This means that the selection probabilities

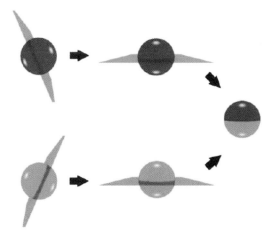

Figure 2.13 Schematic representation of the real-space cut-and-paste crossover operation of Deaven and Ho. (From Ref. [21], with permission from Royal Society of Chemistry.)

p_i in Equation 2.5 do not have to be given by the Boltzmann distribution, especially since the number of structures in the pool N_p does not reflect the (unknown) number of states (structures) that are relevant at the crossover temperature T_c. Based on computational resource considerations, N_p is chosen to be on the order of tens or hundreds; thus, as already mentioned in the context of differences between Boltzmann selection and roulette-wheel selection, we can have some choices in defining the fitness F_i other than Equation 2.6. There is little advantage in exercising this option other than ensuring that the algorithm is robust with respect to changes in the definition of F_i.

2.4.2
Crossovers and Periodic Boundary Conditions

This section is applicable mainly to 3D crystal structure prediction, but brief references are made to other relevant problems as well; it is not at all applicable to cluster structures (0D). In the case of crystal structures and surface slabs, the cut-and-paste operation using a plane does not create only one interface (along the cutting plane) between the parent structures. This is because for surfaces and crystals, there are periodic boundary conditions, and these are not necessarily replicating the cutting plane along itself. Abraham and Probert [19] introduced the idea of cutting operations that are compatible with periodic boundary conditions and have shown that at least in several cases such operations lead to faster convergence than the planar crossovers. Furthermore, in order for crossover operations to be possible between parent structures with different lattice vectors, the mating operation has to be done in the fractional coordinate space. In these fractional coordinates, all structures have the shape of adimensional unit cubes: therefore, they can be cut and pasted in any way to produce a child structure that is also an adimensional unit cube. Using a randomly oriented plane as in the case of clusters will increase the number of interfaces between the parents, with one interface at the cutting plane and others on portions of the boundaries of the simulation cell. Using two periodic functions that are compatible with the periodic boundary conditions leads to exactly two interfaces, that is, the cutting surfaces themselves (Figure 2.14).

Often, for the sake of simplicity of implementation, the cuts involved in the crossover are not done by two periodic or sinusoidal functions, but rather by two or more planes parallel to one another and to one of the faces of the unit cube in fractional coordinates – not always the same face and not with the same interplanar separation. In this sense, the crossover operations resemble those in the binary space even more. Therefore, we can now summarize the steps of a crossover:

i) Transform each parent structure to polar coordinates.
ii) Perform crossover on the common shape (unit cube), which also becomes the shape of the unit cell of the crystal child in fractional coordinates.
iii) Assign physical lengths and angles to the child structure.
iv) Apply, with a small but prescribed, nonzero probability, a mutation to the child structure.

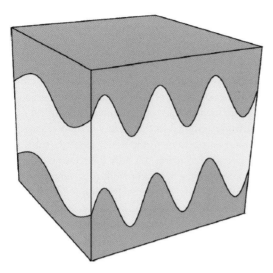

Figure 2.14 Real-space representation of the periodic cuts in the crossover operation proposed by Abraham and Probert [19]. The cuts are calculated in fractional coordinates that allow crossover between parents with differently shaped cells. The dark and light gray sections represent parts of the child that come from the first and second parent, respectively. (Adapted from Ref. [19], with permission from American Physical Society.)

v) Relax the child structure (potentially, but not necessarily mutated), using relaxation algorithms that change the shape and volume of the periodic cell in addition to the atomic coordinates.

The last step is specific to the 2D or 3D crystal structures, but does not apply to surface reconstructions (2D) problem since the surface unit cell is dictated by the surface orientation. Steps (i) and (ii) involve a straightforward vector transformation as explained at length in several references [19,22]. The location of an atom with the position vector **r** can be expressed in terms of fractional coordinates $\eta_j, j = 1, 2, 3$ as

$$\mathbf{r} = \eta_1 \mathbf{a} + \eta_2 \mathbf{b} + \eta_3 \mathbf{c} = \mathbf{A}\eta, \tag{2.8}$$

where $0 \leq \eta_j < 1$ ($j = 1, 2, 3$), **a**, **b**, and **c** are the periodic vectors of the supercell, and the matrix **A** is formed by arranging the periodic vectors as columns (i.e., the first column of **A** contains the Cartesian components of **a**, etc.). The column vector η of components $\eta_j, j = 1, 2, 3$, can be extracted from Equation 2.8 via matrix inversion, $\eta = \mathbf{A}^{-1}\mathbf{r}$. Once in the fractional coordinate space (η space), the cut-and-paste operations between parents are straightforward.

Step (iii) requires some discussion: After the child is created via crossover operations, what should its shape and dimensions be? It is customary in the literature to define the periodic vectors of the child structure as some weighted average of the periodic vectors of the parent structures. If a_A, b_A, c_A and a_B, b_B, c_B are the periodic vectors of the two parent structures A and B, respectively, we can define the first periodic vector of the child as

$$a = \frac{n_A a_A + n_B a_B}{n}, \qquad (2.9)$$

where n_A (n_B) is the number of atoms from parent A (B) that end up in the child structure, and $n = n_A + n_B$ is the number of atoms of the child. The other periodic vectors of the child structure will be defined in the same manner as Equation 2.9. With the physical shape and dimensions assigned to the child, one may decide to apply, with a small probability, a mutation operator to the newly created structure [step (iv)]. Such mutations are described in Section 2.5 and their purpose is to maintain diversity in the genetic pool. Thereafter, one can readily proceed to step (v) in which local minimum relaxation of the periodic box and of the atomic coordinates is simultaneously done; such relaxation procedures are routinely implemented in various software packages. Recalling the flow of a typical genetic algorithm (Figure 2.11), we have the option to produce more than one child structure at a time ($N_c \geq 1$). In this case, relaxation of all child structures considered should be done simultaneously, in a parallel manner.

We caution that even if the parent structures have the same number of atoms and composition, the child structure does not necessarily end up with this same number of atoms or composition. One can choose to enforce constant number of atoms or to allow this number to vary within a reasonable range. So far, the latter option has been seen to result in somewhat faster convergence for finding surface reconstructions [15] and for 3D crystal structure predictions [19]. Still, the former option is necessary if fixed-composition GA runs are desired for crystalline compounds or alloys; in this case, if the newly created structure does not have the nominal composition, it is discarded prior to any energy or fitness calculation.

Extension of this crossover operation formalism to pure 2D and 1D crystals is straightforward, with the note that the 2D and 1D structures have to be placed in the same locations of the parent supercells, otherwise we end up with child structures that can be broken or disjointed by the cut-and-paste operations (we do not have to worry about this in 3D because there is no vacuum space in any of the periodic directions).

2.5
Mutations

Historically, mutations are introduced to change structure in ways that are not readily allowed by crossovers, with the purpose of maintaining a certain level of diversity into the genetic pool. During the algorithm, we want to distinguish between zero-penalty mutations and regular mutations. While any of these mutations can be applied at various stages in the evolution of a GA run, in our own work we tend to apply zero-penalty mutations to parent structures before a crossover is performed and the regular mutations to child structures created after crossover [i.e., at step (iv) in the procedure of Section 2.4.2].

2.5.1
Zero-Penalty Mutations

We call zero-penalty mutations those operations that do not change the energy, or cost, or fitness functions of a structure. Therefore, they do not require a relaxation into the local energy minimum lying closest to their configuration, so there is nearly no computational cost involved in performing them. These mutations are generally symmetry operations that leave the physical structure and energy intact. For the case of clusters, let us imagine two identical parent structures whose atoms have the same absolute Cartesian coordinates in a given reference frame. If we perform any planar cut-and-paste operation on these two identical parents, we end up with a child structure that is identical to them, therefore this crossover amounts to no progress. However, if we apply a zero-penalty mutation by performing random spatial rotations on each of the parent structures prior to the crossover, the resulting child will not be identical to either of the two parents. While this is not always explicitly described in the literature, it is often implicitly applied: see, for example, Figure 2.13 in which the two parent clusters have been rotated prior to splicing.

In the case of crystal 2D and 3D crystal structures, these zero-penalty mutations are random spatial translations, which obviously do not change the energy of the structures or the size and shape of their periodic cells. In the case of 1D nanostructures, we can have arbitrary translations along the periodic direction, but also arbitration rotations about the axis of the wire and an 180° rotation about an axis perpendicular to the wire axis. As mentioned, the zero-penalty mutations are extremely versatile: even if we start from a generation zero that is made of identical structures (which not many reports in the current literature advise), performing a randomized zero-penalty mutation on at least one parent structure prior to entering it in the crossover will immediately create diversity and generate nontrivial child structures.

2.5.2
Regular Mutations

These are mutations performed on the child structures in order to help the GA cross new energetic barriers and explore a somewhat wider region of configurational space. The regular mutations in real space can be either small displacements of a prescribed maximal magnitude applied to one or more randomly chosen atoms, or swaps of atoms of two different types applied to one or more pairs of atoms. Since the regular mutations require relaxation of the structure and therefore consume computational resources, their application should be done judiciously, that is, with a reasonably small probability.

At present, there are no particularly convincing arguments for how often to perform regular mutations on a newly created child structure, but there is sufficient evidence that mutations affect the GA progress in two ways: they can make the progress slower, and they can help find the global minimum. Empirically, we have noticed the following fact: If the crossover operation that we design for a certain problem is sufficiently versatile in exploring configurational space, then the use of mutations is obviated. On the other hand, if the crossover operation is poor (e.g., a crossover based on a diagonal-only cut in the determination of surface reconstructions), then no reasonable amount of regular mutations can bring the GA to find the ground state. To illustrate this point clearly, consider the first cut-and-paste operations of Deaven and Ho, operations that use a cutting plane crossing always through the center of the cluster. Within a given time, the use of this crossover does not find the global minimum for C_{20}, but the algorithm gets trapped into monocyclic carbon rings [1]; introducing mutations with a probability of 5% leads the GA run to finding polycyclic caps, with lower energies than the monocyclic rings (Figure 2.15).

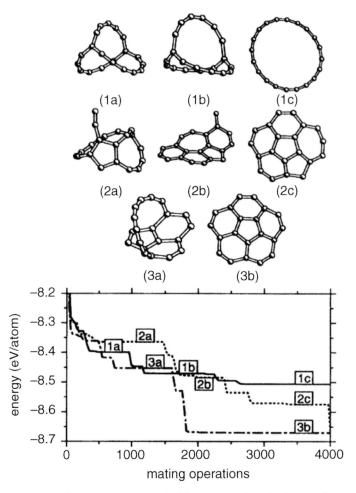

Figure 2.15 Genetic algorithm search of the structure of C_{20} for the case of no mutations (solid line) and for two cases where mutations are applied with probability of 5% (dashed and dot-dashed lines). (From Ref. [1], with permission from American Physical Society.)

If the crossover operation was "richer," for example, by allowing the plane to cut a distance d away from the center and thus to mix a larger fraction of one parent with a smaller fraction from the other, then for certain distances of the plane from the center (those close to the cluster radius) the crossover essentially amounts to mutation since only a very small part of a parent is changed. Indeed, our tests show that with this "richer" crossover, the ground-state structure for carbon clusters with 20, 30, and 60 atoms are always found without introducing any regular mutations.

2.6
Updating the Genetic Pool: Survival of the Fittest

The next important step in the GA flowchart (Figure 2.11) is the updating of the genetic pool. Sometimes this step is called survival of the fittest and we adopt this nomenclature here, but we often see in the literature the selection for mating being called survival; this is probably because at the selection step it is decided which member structures of the population are selected for the mating pool, thus having the chance to create offspring and having their "genes" propagate/survive. We define here a generation as the number N_c of simultaneous crossovers performed between members of N_m of the mating pool (refer to Figure 2.11). This number is a user choice, and we can choose this number of crossovers that defines a generation to be some fraction α of N_m, $N_c = \alpha N_m$. If a crossover results in an unacceptable number of atoms or composition (as defined by falling out of a prescribed range) or if it results in some unacceptable "hard" geometrical criterion (as discussed in Chapter 3), we can convene to not count it against the prescribed number of crossovers N_c, and repeat it until the number of atoms, composition, and/or geometry are acceptable for the requirements of the optimization problem. This convention makes sense, because relaxation is the time-consuming step and it is not done if the number of atoms or concentration for a child structure is unsatisfactory. With the newly created N_c structures, how do we evolve the genetic pool to the new generation? The straightforward way to do it is to include these children in the population if their energy (or cost function or fitness function values) is better than that of some of the existing population. Procedurally, we should

i) rank-order the old N_p structures and the newly created N_c structures together in a larger set (called reunion in Figure 2.11) comprising $N_c + N_p$ relaxed structures; and
ii) keep the fittest N_p structures from this larger set and discard the rest.

At this point the pool is updated, and we say a generation has passed and we increase its generation index by 1. Some of the newly created structures may enter the genetic pool, whereas the least fit N_c structures (those with high energies) will be removed regardless of whether they are newly created or not: this is why this particular stage of the GA is sometimes called the survival of the fittest. In practical implementations of GA, there is often (but not always) another criterion for including a structure into the population: not only should the child be sufficiently fit or sufficiently energetically favorable, but it should also be structurally different from any other member in the pool. Imposing this requirement is usually done by declaring identical structures whose formation energy f are within a small tolerance δf from one another; the value of this tolerance requires certain numerical experimentation and may change from one problem to another. In 3D, other ways to assess if structures are identical or not are discussed in Section 3.5.

Based on practice, sensible relations between N_p (number of structures in the genetic pool), N_m (number of structures selected for mating), and N_c (number of crossovers that define a generation) may be

$$N_m = \frac{N_p}{2} = 2N_c, \qquad (2.10)$$

with the size of the pool N_p being between 5 and 100. As we shall see in the following chapters, there are many examples of updating the pool in different ways than the one just described, with similar degrees of success. There is little or no advantage in trying to isolate one or another, it is sufficient to say that both diversity of the population and the computational resources have to be managed well, and that such management usually involves a certain amount of trial and error by the user.

2.7
Stopping Criteria and Subsequent Analysis

The algorithm slows down after a while, in the sense that there will be increasingly fewer new structures accepted in the genetic pool as generations pass. This happens when most structures in the pool have already achieved configurations with satisfactorily low energies and it has become hard to do better: the pool has evolved, and the algorithm can stop. The most common stopping criteria are as follows:

i) Reaching a prescribed total number of generations.
ii) Reaching a desired number of generations past the point where the best structure in the pool has changed for the last time.
iii) Reaching a predefined energy tolerance for the *average* energy (or any cost function in use).
iv) When the average pool energy is within a prescribed tolerance from the lowest energy in the pool.

The analysis that usually follows is in the form of plotting the history of the evolution in some way that makes it readily obvious that there are no more changes in the best structure reached or in, for example, the average energy per atom across the pool. This could be the energy of the best structure in the pool as a function of time, or the average energy as a function of time, or the entire energy spectrum of cost functions at every generation. Analysis does not necessarily occur only in the way of checking if the algorithm has done its job, but also in ways in which it reveals some fundamental information about the PES landscape of the system (Section 3.5).

References

1 Deaven, D.M. and Ho, K.M. (1995) *Phys. Rev. Lett.*, **75**, 288.
2 Kroto, H.W., Heath, J.R., O'Brien, S.C., Curl, R.F., and Smalley, R.E. (1985) *Nature*, **318**, 162.
3 Kosimov, D.P., Dzhurakhalov, A.A., and Peeters, F.M. (2010) *Phys. Rev. B*, **81**, 195414.
4 Brenner, D.W., Shenderova, O.A., Harrison, J.A., Stuart, S.J., Ni, B., and Sinnot, S.B. (2002) *J. Phys.: Condens. Matter*, **14**, 783.
5 Cai, W., Shao, N., Shao, X., and Pan, Z. (2004) *J. Mol. Struct.: THEOCHEM*, **678**, 113.
6 Wang, C.Z. and Ho, K.M. (2004) *J. Comput. Theor. Nanosci.*, **1**, 3.
7 Wang, Y., Zhuang, J., and Ning, X.J. (2008) *Phys. Rev. E*, **78**, 026708.
8 Ho, K.M., Shvartsburg, A.A., Pan, B., Lu, Z.Y., Wang, C.Z., Wacker, J.G., Fye, J.L., and Jarrold, M.F. (1998) *Nature*, **392**, 582.
9 Qin, W., Lu, W.C., Zhao, L.Z., Zang, Q.J., Chen, G.J., Wang, C.Z., and Ho, K.M. (2009) *J. Chem. Phys*, **131**, 124507.
10 Lu, Z.Y., Wang, C.Z., and Ho, K.M. (2000) *Phys. Rev. B*, **61**, 2329.
11 Kresse, G. and Furthmüller, J. (1996) *Comput. Mater. Sci.*, **6**, 15.
12 Plimpton, S. (1995) *J. Comput. Phys.*, **117**, 1.
13 Oganov, A.R. and Glass, C.W. (2006) *J. Chem. Phys.*, **124**, 244704.
14 Wu, S., Umemoto, K., Ji, M., Wang, C.Z., Ho, K.M., and Wentzcovitch, R.M. (2011) *Phys. Rev. B*, **124**, 184102.
15 Chuang, F.C., Ciobanu, C.V., Shenoy, V.B., Wang, C.Z., and Ho, K.M. (2004) *Surf. Sci. Lett.*, **573**, L375.
16 Ciobanu, C.V. and Predescu, C. (2004) *Phys. Rev. B*, **70**, 085321.
17 Lenosky, T.J., Sadingh, B., Alonso, E., Bulatov, V.V., Diaz de la Rubia, T., Kim, J., Voter, A.F., and Kress, J.D. (2000) *Model. Simul. Mater. Sci. Eng.*, **8**, 825.
18 Mo, Y.W., Savage, D.E., Schwartzentruber, B.S., and Lagally, M.G. (1990) *Phys. Rev. Lett.*, **65**, 1020.
19 Abraham, N.L. and Probert, M.I.J. (2006) *Phys. Rev. B*, **65**, 224104.
20 Darby, S., Mortimer-Jones, T.V., Johnston, R.L., and Roberts, C. (2002) *J. Chem. Phys.*, **116**, 1536.
21 Johnston, R.L. (2003) *Dalton Trans.*, 4193.
22 Trimarchi, G. and Zunger, A. (2007) *Phys. Rev. B*, **75**, 104113.

3
Crystal Structure Prediction

This chapter addresses in more detail the problem of crystal structure prediction using genetic algorithm (GA) within the real-space representation that we have discussed in the previous chapters. As mentioned in Chapter 1, in our view the main foundational works are as follows:

i) The pioneering work by Deaven and Ho [1] in which the authors put forth the cut-and-paste operations in real space and have shown that systems with $N \sim 60$ atoms are solved successfully. Although originally applied to clusters only, these operations have been refined over time by other authors and also extended to fully periodic crystal structures [2].
ii) The work of Abraham and Probert [2], who have recognized the versatility of performing mating operations in fractional coordinates, which have enabled the crossover between parent structures having different shapes and sizes.
iii) The works of Oganov and Glass [3] and Trimarchi and Zunger [4], who assessed the difficulty of the problem, developed a suite of genetic operations, combined GA with DFT (density functional theory) relaxations, and showed that a tractable number of DFT structural relaxations (usually several hundred) are sufficient to predict the global minimum structure for certain elemental and compound systems. Furthermore, both groups have also shown that the stable compositions of an alloy and the corresponding atomic structures can be determined simultaneously in GA runs with variable composition across the entire composition domain [5,6].

This chapter consists of two parts: in the first part, we describe certain issues that have to be considered in order to craft a powerful genetic algorithm for crystal structure prediction. These include an assessment of the complexity of the problem (Section 3.1), coupling of the algorithm with empirical potentials, or DFT calculations, or both (Section 3.3), improving efficiency by imposing certain type of conditions during the run (Section 3.2), and assessing and maintaining the structural diversity of the pool (Section 3.5). With the exception of the discussion on atomic interactions – empirical potentials continuously fitted [7] to DFT calculations (Section 3.3.3), or followed by DFT (Section 3.3.1) – the main ideas pertaining to the development of the algorithm that are covered in this chapter have been put forth in the works of Oganov and coworkers, in particular in Refs [3,6,8,9]. The transition between algorithm development issues and the GA applications in crystal structure prediction is provided by a section in which we discuss the variable-composition version [5,6] of the algorithm (Section 3.6).

In terms of applications, we have investigated the literature and have found that Oganov and coworkers so far lead the way in terms of applications and discoveries based on GA for structure prediction. Since various accounts of this work have already been published, we do not attempt to review here but do find it very instructive to point out to the reader a few of the successes of GA in terms of discovery of materials with unsuspected properties:

- Ionic high-pressure phase of boron [10]. A new phase of elemental boron was predicted, stable between 19 and 89 GPa, consisting of icosahedral clusters and atom pairs in a lattice akin to that of NaCl.
- Transparent dense sodium [11]. At pressure of about 200 GPa, there is a pressure-induced transformation of sodium (metal) into an optically transparent phase (wide-gap dielectric).
- High-pressure phases of $CaCO_3$ [12]. Using GA, the structure of this mineral was identified – $CaCO_3$ is believed to be the source of significant carbon content in Earth's mantle.
- A superhard monoclinic polymorph of carbon [13]. This newly discovered phase is more stable than graphite at pressures greater than 13 GPa.
- Superconducting high-pressure phase of germane [14].

As examples of predictions of crystal structures and properties using GA, we are including in this chapter recent applications from the Ames Laboratory–University of Minnesota collaboration (Wang–Ho–Wentzcovitch). The three applications from Ames Laboratory and University of Minnesota concern the post-pyrite phase transformation in silica (Section 3.7.1), the high-pressure phases of ice (Section 3.7.2), and the structure and magnetic properties of Fe–Co alloys (Section 3.7.3). To make this chapter self-contained for the possible benefit of graduate students, we described these applications fully in terms of computational details, results, and analysis – an approach that we adopt for the subsequent chapters as well.

3.1
Complexity of the Energy Landscape

To evaluate the complexity of the problem faced with the global structural optimization or prediction of crystal structure, we can start by assessing the number of points on the potential energy surface as [3]

$$C = \binom{n_{\text{grid}}}{n} \prod_i \binom{n}{n_i}, \tag{3.1}$$

where n_{grid} is the number of grid points in which we can sensibly discretize the periodic cell in which we have n total atoms, and n_i is the number of atoms of type i in the computational cell. In terms of the number of grid points, which can be defined simply as $n_{grid} = V/v$, that is, the ratio between the volume V of the cell and an "excluded volume" v that can be occupied by at most one atom; a reasonable value for v is $1\,\text{Å}^3$ [3,4]. Using Sterling's approximation (i.e., the factorials in the combinations appearing in Equation 3.1 expressed as exponentials) and considering the dimensionality of the system

$$d = 3n + 3, \tag{3.2}$$

we can rewrite Equation 3.1 as a simple exponential:

$$C = \exp \alpha d, \tag{3.3}$$

where α depends on the system under consideration. Although the number of combinations of atomic coordinates C is finite, it takes astonishing values even for elemental systems: for 10, 20, and 30 atoms per simulation cell, we have $C \sim 10^{14}$, 10^{25}, and 10^{39}, respectively [10]. Furthermore, the values of C for binary compounds AB are three to eight orders of magnitude larger! These estimations show clearly that the point-by-point energy calculations of the configuration space are not tractable except for maybe the simplest systems.

Expectedly, when local optimization (relaxation into the nearest local minimum) is used, then the space to explore becomes significantly smaller. During relaxation, distances between atoms adjust to meaningful values, all usually higher than $1\,\text{Å}$, and bond angles adjust to those in real structures and in the process decrease the number of possible configurations because not all the combinations in Equation 3.1 will be allowed by the relaxation. As shown in Figure 3.1, the potential energy landscape will now consist in the set of local minima, which are significantly fewer in number than number of combinations C estimated above. The dimensionality of the energy landscape $d = 3N + 3$ becomes significantly smaller: this is denoted by

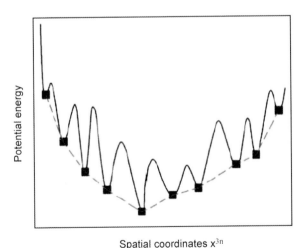

Spatial coordinates x^{3n}

Figure 3.1 Schematic depiction of the potential energy landscape (solid curve) and the reduced landscape obtained through relaxations (dashed line joining local minima). (Adapted from Ref. [9], with permission from American Chemical Society.)

d^* and is smaller than d by the number κ of degrees of freedom that are correlated through structural relaxations [9]:

$$d^* = d - \kappa, \tag{3.4}$$

and the number of combinations becomes much smaller, $C^* \sim \exp(\beta d^*) \ll C$. In the case of Au_8Pd_4, $d^* \sim 11$ ($d = 39$), or for $Mg_{16}O_{16}$ we have $d^* \sim 12$ ($d = 99$). Because of this reduction in the effective dimensionality of the system, at this time GA can reasonably tackle systems with $N < 100$ atoms in the simulation cell.

3.2
Improving the Efficiency of GA

The main way in which the efficiency of GA coupled directly with DFT is presently controlled is by judicious decisions concerning which of the new (children) structures are going to be relaxed. Structural relaxations with DFT are very expensive, each one necessitating on the order of 50–200 ionic steps. This number depends on how far from equilibrium the unrelaxed child structure is: the larger the initial forces on atoms, the longer the ionic relaxation. In turn, each ionic relaxation requires on the order of 20–60 electronic relaxation steps, again with the larger number corresponding to structures farther from equilibrium. Given these numbers, and the fact that a typical DFT ionic relaxation (e.g., using VASP [15]) can require 32–128 processors for 0.5–2 days, we clearly see the need to decide carefully which new structures should be relaxed and which should be discarded without any expensive computation. This means that only structures obeying a set of hard constraints [3] should be relaxed. A reasonable set of such hard constraints that act as gatekeepers against wasteful relaxations [3] are as follows:

i) minimum acceptable interatomic distances;
ii) minimum and maximum ranges of any of the three periodic lengths;
iii) minimum and maximum angles between the periodic vectors.

The minimum interatomic distance should be decided based on whether the GA optimization occurs under applied pressure to the simulation cell, and is usually 0.7 Å; larger values are allowed for calculations in the absence of applied pressure, and furthermore, this minimum interatomic distance can be set separately for each different pair of atomic species. The angles of the computational cell can be safely restricted between any two vectors of the cell falling between 60° and 120°. This restriction on the angular range can be imposed in two ways: one in which we reject any child structure with angles outside the range prior to any relaxation, and another in which we replace the longer vector in a pair making an angle outside the (60°, 120°) range by [8]

$$\mathbf{a}_{new} = \mathbf{a} - \mathbf{b}\,\text{ceil}\left(\frac{|\mathbf{a}\cdot\mathbf{b}|}{|\mathbf{b}|}\right)\text{sign}(\mathbf{a}\cdot\mathbf{b}). \tag{3.5}$$

Reasonable values for all of these hard constraints are given in Ref. [3]. A minimum periodic length, although it is usually imposed as a separate hard constraint, is actually related to the minimum interatomic distance, while the maximum value of the periodic length is indirectly determined by the available resources (and, in this sense, is related to the total number of atoms n).

As a word of caution for possible beginners in the workings of GA, we point out that in the literature the word "constraint" is sometimes also used to mean a restriction of the complexity of the problem in order to make the problem simpler and more tractable. For example, one can fix at least *one* of the following and optimize the rest: periodic lattice vectors, number of atoms in the simulation cell (set to, for example, a very small value), or composition of the simulation cell. In practice, in addition to the hard constraints described above, the only thing that makes sense to restrict from above (though not necessarily fix to a given value) is the total number of atoms because this number is directly linked to the available computational resources for GA investigation.

3.3
Interaction Models

3.3.1
Classical Potentials

In Chapter 2, the general scheme of a GA was described without particular reference to a model for atomic interactions or another. In general, the interaction model can modify the effective dimensionality of the system: this is because the reduced potential energy surface is the set of local minima and this set of minima can certainly be different depending on the atomic interactions used. Unfortunately, there is no way to tell how exactly the number of minima increases or decreases for a given system when the atomic interaction model is changed. Not all available empirical models describe the actual interactions very well, but in some cases (e.g., elemental silicon) such empirical models are sufficiently good (Stillinger–Weber [16], Tersoff [17], Lenosky *et al.* [18], etc.). In these cases, it makes sense for the GA optimization to be carried out using the empirical potential. After the GA run has been carried out with an empirical potential, however, we should use more accurate description of the interactions (such as DFT or tight binding) to recalculate the cost functions of the structure obtained. If the rank ordering of the more energetically favorable structures in the pool does not change too significantly upon recalculation at the level of DFT, then this means that the potential was sufficiently reliable to start with; therefore, after the DFT reordering of the structures, we can have a high degree of confidence in the refined results. Such approach, for example, was taken for the surface reconstructions and nanowire cross-sectional shapes that are discussed in Chapter 5.

3.3.2
Ab Initio Methods

In other cases, especially in case of compounds made of two or more atomic species, the interactions may not be sufficiently well described by any empirical model except for a small set of structures to which the potential was originally fitted; in these cases, algorithms for structure prediction that employ such atomic interaction model are very likely to lead to unrealistic or unphysical structures because of the artifacts of the potential. It is not an easy task to determine a new potential model, or even to reparameterize an old one. In these cases, starting with the works of Abraham and Probert [2], Oganov and Glass [3], and Trimarchi and Zunger [4], workers have made direct use of the DFT methods as ways to determine the cost function. As we can recall, during a GA run the newly created structures (offspring, or children) can be rejected if their formation energy is too high – which is very time consuming because we end up discarding a structure that has already taken considerable resources. Nevertheless, this approach has been very successful, owing in principle to a set of decisions that are taken before proceeding with the DFT-level relaxations (see Section 3.2). In Section 3.7.1, we will see an example of DFT-based GA in which the high-pressure phases of silica are determined and discussed.

3.3.3
Adaptive Classical Potentials

The very selective decisions to proceed with the DFT relaxations are a powerful way to ensure that the algorithm is not only tractable, but also very efficient. Another way toward increasing efficiency is to use, during the algorithm, a combination of the empirical potential and DFT in a way in which the parameters of the empirical potential are continuously fit to DFT structures and forces or energies. This way, because of the requirement that the parameters of the potential "adapt" to DFT values continuously, overreliance on the (possibly artificial) starting values of the interatomic potential parameters is dramatically reduced. Figure 3.2 shows schematically how a GA with adaptive empirical potential works.

The flowchart of a typical GA run with adaptive empirical potential has two main parts:

i) a GA optimization based solely on the empirical potential; as such, this is very similar to the flowchart in Figure 2.11;
ii) the second part, in which the parameters of the potential are being recalculated to DFT values for structural parameters and forces.

Upon meeting the exit criterion from the first loop, the algorithm enters the fitting loop, in which the most favorable N_{DFT} structures from the set of N_p are selected for *single-point* DFT total energy calculations. These structures are not equilibrium structures at the DFT level, and therefore atoms will have forces on them that are not relaxed (given that the calculation is single-point). Using the DFT-calculated energies and atomic forces for these structures (which have been optimized at the

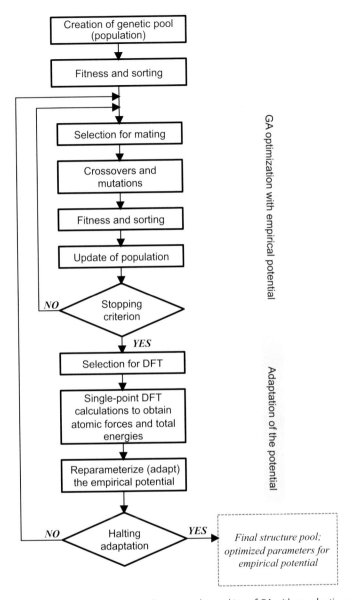

Figure 3.2 General flowchart illustrating the working of GA with an adaptive empirical potential.

empirical level), we change the parameters of the potential so as to match the energies and forces. At present, this energy-and-force fitting procedure is in place for Lennard-Jones (LJ) potentials [19] and for embedded atom potentials [20]. If the number of such parameter fits reaches a prescribed value (or some other criterion that makes sense), then the algorithm exits. If not, then the newly fit potential parameters are fed back to the first GA loop based on the empirical potential, and the algorithm continues.

There are several key features that make the GA with adaptive empirical potential very efficient, and potentially applicable to very large systems (hopefully, past the current limit of 100–200 atoms). First, at no point in run we would have to do costly DFT relaxation of any newly created children structures with significant stresses in them that appear due to cut-and-paste operations. In fact, there is no DFT relaxation at all, because the refitting of the potential parameters is done based on forces and energies computed in single-shot calculations for a number N_{DFT} of structures (i.e., only electronic relaxation and no ionic relaxation). Usually, it makes sense to pick $N_{DFT} = N_p/2$, that is, to use in the fit the best half of the genetic pool. Second, the genetic algorithm carried out with the empirical potential (either with original parameter values or with refit ones) is very rapid. Third, the fact that the fitting is done on the best structures resulting from GA coupled with the empirical potential ensures that the algorithm not only finds crystal structures but also eventually converges on a set of parameters. At present, the only drawback seems to be that we do not have yet a sufficient number of starting potentials. We will see an application of this GA based on adaptive empirical potential in Section 3.7.2.

3.4
Creating the Generation-Zero Structures

The power of GA for crystal structure prediction, especially when coupled directly with DFT total energy calculations, consists in the following:

a) It can predict the structure without any experimental input. However, one can introduce (seed) known structure in the genetic pool to act as a "mark" against which the pool should improve as the algorithm progresses.
b) It can predict the structure in regimes of pressure that are not currently achievable or directly measurable experimentally (such as the pressure in the Earth's mantle).
c) It alleviates significantly the quest for materials design, in which we have to first uncover the new structures and properties in certain regime of conditions (e.g., high pressure) before attempting to design materials for these conditions.

This being said, how does one start the algorithm? What is the generation zero (refer to Figures 2.11 and 3.2)? There are several ways to do that:

Random initial pool. It is relatively easy to create sets of structures in which the locations of the atoms are arbitrary. Still, even in this case, one should ensure that the hard constraints defined above (Section 3.2) are obeyed – for example, by discarding structures until all members of the pool obey the hard constraints. Otherwise, the algorithm will be left to optimize, using DFT, structures that are too far away from equilibrium. This type of initialization poses a problem for systems with very large numbers of atoms because of the very large configurational space.

Random pool seeded with reasonable structures. To alleviate the problem of too large configurational spaces sampled inadequately by random structures in generation zero, we can introduce from the beginning in the pool known structures of the material or compound under consideration. If these known structures are sufficiently many, then it is very likely that some of the structural traits present in them will make the algorithm progress faster.

Growth from single particle of each type. This way of initialization of the pool is somewhat extreme, in that many DFT relaxations should be performed until the material reaches a density sufficiently high and compatible with the applied external pressure. In case of the use of empirical potential, this is not a problem and in fact it is an extremely versatile way to check that the ground-state structure has been obtained from runs initialized in different ways.

In principle, the diversity of the structures in the first generation should be important for the success of the algorithm. However, if one allows for zero-penalty mutations of the parents before mating, the diversity is "refreshed" before each crossover and the algorithm can still progress significantly even when starting with *identical* member structures in generation zero! There is a subtle point here to note and that is the link between the initialization of the pool and the update: are the updates to the genetic pool done in such a way as to prevent the duplication of the structures in the pool? If the answer is yes, then the initialization becomes less important especially if zero-penalty mutations are in play.

3.5
Assessing Structural Diversity of the Pool

3.5.1
Fingerprint Functions

In discussing the diversity of the pool, we should note that there should be a balance between the diversity allowed or required and the efficiency of the algorithm. For example, if we insist that diversity should be explored thoroughly, then the algorithm might require an extremely long time to converge; on the other hand, if we are not mindful of allowing or encouraging diversity in the structures, then the algorithm converges prematurely into a false global minimum. While diversity can be ensured qualitatively through the initialization of the pool and/or through the zero-penalty mutations (defined in the previous chapter), it is still important to have actual quantitative measures of it. This need has been recognized by Oganov and Valle [21], who introduced, among others, the following measures for characterization of structures and of the genetic pool:

a) *Fingerprint functions.* These functions represent a measure of the individuality of each configuration, and therefore should not depend on anything else or be deduced from anything else but the structure itself. Any definitions should be sensitive to the degree of ordering, coordination numbers, and symmetries of the crystal, should hopefully be related to some experimentally accessible quantity, and should be robust against numerical noise [21]. There are several definitions available for such functions, but the one that satisfies all these conditions and seems most natural to adopt, at least at present, is

$$\mathcal{F}^{(ij)}(r) = g_{ij}(r) - 1, \tag{3.6}$$

where $g_{ij}(r)$ is the radial distribution function (a.k.a. two-particle correlation function) for the species i and j. The subtraction of 1 in the previous equation is done so that the fingerprint function vanishes in the limit $r \to \infty$. Therefore, with this definition, we do not have a fingerprint function (except for the case of single-species systems) but there is a set of such functions – one fingerprint function for each pair of atomic species. As an example, Figure 3.3 shows the fingerprint of the ground state of the Au_8Pd_4 system with 12 atoms in the computational cell.

b) *Distance between structures.* Given two structures, we would need a quantitative way to determine how different they are from one another. This metric can also be defined in several ways, but we will give only one such metric here. First, the fingerprint functions that characterize each of the two structures whose distance is calculated are discretized over the r space, and thus become "vectors" $F = \{F_k^{(ij)}, i,j = 1, 2, \ldots, n_{\text{spec}}; k = 1, 2, \ldots, k_{\text{max}}\}$ with k being the discretization index that takes values from $k = 1$ (when $0 \leq r < \Delta r$) to $k = k_{\text{max}}$ (when $r_{\text{max}} - \Delta r \leq r \leq r_{\text{max}}$):

$$\mathcal{F}_k^{(ij)} = \mathcal{F}^{(ij)}(r = k\Delta r). \tag{3.7}$$

Then the distance between the structures is defined as [21]

$$D_{12} = \frac{1}{2}\left(1 - \frac{\mathbf{F}_1 \cdot \mathbf{F}_2}{|\mathbf{F}_1||\mathbf{F}_2|}\right). \tag{3.8}$$

This is often called the cosine metric, to distinguish it from other metrics – for example, from the regular Euclidian distance. The dot product in Equation 3.8 should be computed in a way that takes into account the many fingerprint

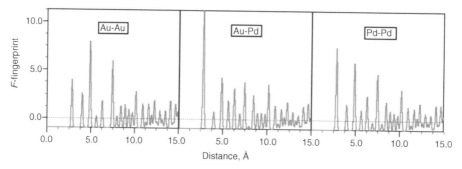

Figure 3.3 The components of the fingerprint function F for the ground-state structure of the Au_8Pd_4 system. (From Ref. [21], with permission from American Institute of Physics.)

functions generated when the number of atomic species (n_{spec}) is greater than 1:

$$\mathbf{F}_1 \cdot \mathbf{F}_2 \equiv \sum_{i=1}^{n_{\text{spec}}} \sum_{j=1}^{n_{\text{spec}}} \sum_{k=1}^{k_{\text{max}}} \mathcal{F}_{1,k}^{(ij)} \mathcal{F}_{2,k}^{(ij)}. \tag{3.9}$$

With this definition of the dot product, the norm of a vector $\mathbf{F}$ (such as those that appear in the denominator in Equation 3.8) is

$$|\mathbf{F}| \equiv \left(\sum_{i=1}^{n_{\text{spec}}} \sum_{j=1}^{n_{\text{spec}}} \sum_{k=1}^{k_{\text{max}}} (\mathcal{F}_{1,k}^{(ij)})^2 \right)^{1/2}. \tag{3.10}$$

c) *Structural diversity.* With the distance between any two structures in the genetic pool defined as above (Equation 3.8), we can have measure of the overall diversity of the genetic pool. This measure is called quasi-entropy [21] and is defined as

$$S_{\text{pool}} = \langle \ln(1 - D_{pq}) \rangle, \tag{3.11}$$

where the average is taken over all pairs (p, q) of structures in the genetic pool.

This particular metric is useful to critically assess the success of a genetic algorithm: we should not simply monitor the best energy structure and declare convergence when it has not changed for a long time, but should also check the structural diversity of the pool using Equation 3.11. If this quantitative diversity, as given by Equation 3.11, decreased sharply in the first few generations, this most likely means that the algorithm converged prematurely. When using a particular GA code with explicit prevention of the duplicate structures, we can directly verify the effectiveness of this by checking the progress of S_{pool} during the run.

3.5.2
General Features of the PES

In addition to the practical use of fingerprint, distance, and diversity functions (Equations 3.6–3.11) for assessing the progress of a GA run, there is a certain amount of fundamental knowledge that is enabled by these quantitative expressions. This fundamental knowledge refers to the general characteristics of the PES for crystalline or alloy systems. For example, we can figure out if a certain system under consideration has the PES with one "funnel" structure, or with multiple funnels. Even though we deal with the reduced PES (Figure 3.1), its structure is still highly complex and the extraction of chemically relevant information from it (such as metastable states, transition states, and energy barriers) is a challenge. The metrics defined in the previous section can help with this challenge; this task is enabled by the following relation [21] between the relative energy ΔE_V of a structure to the ground state (at constant volume) and its fingerprint $F^{(ij)}(r)$:

$$\Delta E_V = 2\pi \sum_{i=1}^{n_{\text{spec}}} \sum_{j=1}^{n_{\text{spec}}} \int_0^\infty \mathcal{F}^{(ij)}(r) U_{ij}(r) r^2 \, dr, \tag{3.12}$$

where U_{ij} is the interatomic, two-body potential acting between the species i and j.

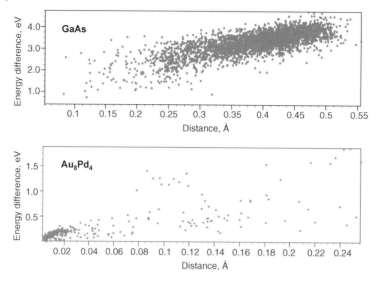

Figure 3.4 Correlation between the relative energy with respect to the ground state and the cosine distance to the ground state, for two different systems, Ga–As (8 atoms per supercell) and Au_8Pd_4 (12 atoms per supercell). The data used in this plot were obtained from GA runs and random samplings. (Adapted from Ref. [21], with permission from American Institute of Physics.)

Equation 3.12 can be used to compute the energy difference to the ground state of various structures in the genetic pool and then plot it against the distance (metric) between that structure and the ground state. Why is this important? Because if there is any clear correlation between the relative energy and the distance, then this would be a direct confirmation of the funnel structure of the PES; it is worth noting that the funnel structure was long confirmed for small clusters, but only recently these PES studies confirm it for crystal structures as well. Figure 3.4 shows such direct correlation for two different systems, using data obtained from GA runs [21].

The single-funnel structure of the PES is a common characteristic for many crystals, but it is by no means general. For example, the propensity of Mg to take many different coordination numbers (from 4 to 8) in oxides and silicates leads to a multifunnel structure of the PES of the corresponding systems, with significant overlap between the funnels. Another situation is one in which the multiple funnels are clearly separated from one another, and this is illustrated for the case of H_2O with 12 atoms per supercell (refer to Ref. [21], and also see Figure 3.5).

3.6
Variable Composition

A major breakthrough in the development of GA for crystal structure prediction has occurred with the realization that the simultaneous prediction of *all* the stable compositions and their corresponding crystal structures is possible [5,6]. There are

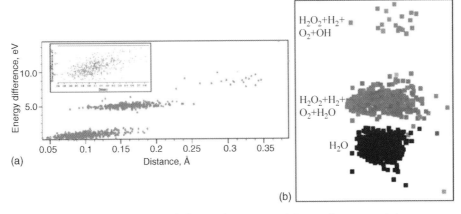

Figure 3.5 Reduced PES of a supercell of H_2O with 12 atoms. (a) Energy–distance correlation. (b) Map of the reduced PES, with the black points indicating the lower-energy structures. (From Ref. [21], with permission from American Institute of Physics.)

several important modifications of the GA at constant compositions that have to be done in order for this feat to be achieved.

First, one has to allow the crossover and mutation operators to act so as to change the composition, with the resulting structures being subjected to DFT relaxations if the hard constraints are satisfied. We can initialize the pool with structures that have a prescribed range of compositions, but this is not always very necessary if the crossovers and mutations are already allowed to change the composition.

Second, we have to redefine our cost function from enthalpy per atom to a function that represents the "distance" to a so-called convex hull – that is, to a convex curve in the enthalpy (per atom)–composition space on which the lowest-enthalpy structures lie for the entire range of compositions. At zero Kelvin and no applied external pressures, the formation energy per atom of a binary structure A_xB_{1-x} is given by Equation 2.2, which we cast here in terms of the concentration x:

$$f(x) = E(A_xB_{1-x})/n - x\mu_1 - (1-x)\mu_2. \tag{3.13}$$

The cost function to optimize now is defined as the distance $\delta(x)$ between the formation energy $f(x)$ and the convex hull [5]. The convex hull $C^{(k)}(x)$ at any time during the GA evolution (say, generation k) is defined as the lowest value of $f(x)$ achieved from the beginning of the GA run up to generation k, for each concentration x. If for some certain concentration x_α, the system is unstable against decomposition into two nearby compositions x_β and x_γ ($x_\beta < x_\alpha < x_\gamma$), then the convex hull value at x_α, at that particular point in time, will be a line joining the two nearby x_β and x_γ points. Formally, the cost function to optimize is the distance to the convex hull [22],

$$\delta^{(k)}(x) = f(x) - C^{(k)}(x), \tag{3.14}$$

where k is the index of the current generation and $C^{(k)}(x)$ is the approximation to the convex hull developed since the beginning of the run until generation k. This continuous optimization of the convex hull is illustrated in Figure 3.6.

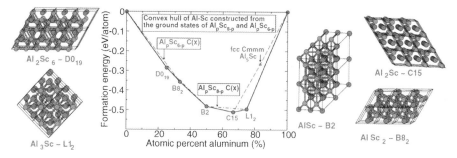

Figure 3.6 Convex hull for the Al–Sc system (solid line), along with the ground-state crystal structure for certain compositions. (From Ref. [5], with permission from American Physical Society.)

As shown by Trimarchi, Freeman, and Zunger, this approach is more versatile than the single-composition GA optimization since a *single* evolutionary search uncovers lowest-energy structures for each stable composition, and also selects from those the ones that represent ground states. Trimarchi *et al.* correctly predict the stable compositions and structure types: AlSc$_2$ (B8$_2$ type), AlSc (B2), and Al$_2$Sc (C15) in periodic supercells with six atoms, and Al$_2$Sc$_6$ (D0$_1$9), AlSc (B2), and Al$_3$Sc (L1$_0$) in the supercells with eight atoms [5].

The extension of this approach to case where external pressure is applied is straightforward and requires the formation energies to be replaced by formation enthalpies throughout the procedure. Figure 3.7 shows the variable-composition results for the Mg–Fe systems subjected to high pressures (350 GPa, similar to those in the Earth's core). Although pure iron has a hexagonal close-packed ground state at this pressure, alloying with Mg leads to the stabilization of bcc structure for many of the Fe–Mg alloys with intermediate compositions.

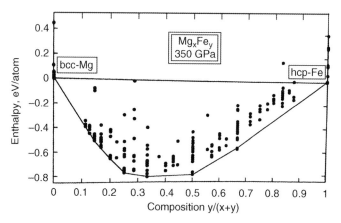

Figure 3.7 The convex hull for the Fe–Mg alloy at 350 GPa. Structures are computed using variable-composition GA coupled with DFT calculations in the generalized gradient approximation [50]. (From Ref. [6], with permission from Wiley-VCH.)

3.7
Examples

3.7.1
Identification of Post-Pyrite Phase Transitions

The discovery of the post-perovskite (PPV) transition of $MgSiO_3$ near Earth's core–mantle boundary conditions [23–25] reminded us that minerals can have truly unexpected high-pressure behavior. This finding naturally introduced a new question: what is the next high-pressure polymorph of $MgSiO_3$? The answer is fundamental for modeling the interiors of recently discovered exoplanets, particularly the terrestrial type [26,27], and the cores of the solar giants, where pressures and temperatures can reach 4 TPa and 21 000 K [28]. In 2006, the dissociation of $MgSiO_3$ PPV into CsCl-type MgO and cotunnite-type SiO_2 was predicted at 1.1 TPa [29]. This prediction was based on the assumption of a sequence of pressure-induced transitions in SiO_2: rutile → $CaCl_2$ → α-PbO_2 → pyrite → cotunnite phases. Experimentally, only phases up to pyrite-type SiO_2 have been observed to date [30]. This sequence of transitions seemed very reasonable because

i) MgF_2, a low-pressure analogue of SiO_2, undergoes the same sequence of pressure-induced transitions from rutile up to pyrite and then transitions to the cotunnite phase [31] preceded by phase X in a very narrow pressure range [32], and
ii) the cotunnite phase has cation coordination number (CN) higher than that of the pyrite phase (9 versus 6).

Nevertheless, it is not guaranteed at all that the cotunnite phase is the real post-pyrite phase of SiO_2. The predicted transition pressure to the cotunnite phase (∼0.69 TPa) [29,33] is still too high to be observed in static compression experiments. Prediction of high-pressure phases in the multi-Mbar regime is a difficult problem and behavior of low-pressure analogues is often invoked. Comparison of enthalpies and/or Gibbs free energies of potential structures is the method frequently used (see Refs [33,34] for post-pyrite SiO_2) to predict phases at ultrahigh pressures. However, these strategies do not guarantee that the true stable structure is identified. Structural search using GA is much more likely to catch truly stable phases and this method has been proven to work very efficiently [3,4]. Here we show that a first-principles GA search predicts the Fe_2P phase as the first post-pyrite phase of SiO_2. The cotunnite phase has very competitive enthalpy but definitely higher than that of the Fe_2P phase beyond the stability field of the pyrite structure. However, the cotunnite phase is stable at high temperatures according to quasiharmonic (QHA) free-energy calculations [35]. These two structures are very closely related, and the existence of another competitive and structurally intermediate phase, the NbCoB type, suggests a gradual crossover between them. We also discuss the effect of these new phase transitions on the dissociation of $MgSiO_3$ PPV.

3.7.1.1 Computational Details

First-principles GA structural searches were performed at 0.5 and 2 TPa. The details of our genetic algorithm were described in Ref. [36]. The number of structures in the GA pool was 32 or 64. The candidate structure pool was initially generated from experimentally known and randomly generated structures. We considered primitive cells with 1.8 SiO_2 formula units (FUs). We used the local density approximation (LDA) [37,38]. Two sets of silicon and oxygen pseudopotentials were generated by Vanderbilt's method [39]. For the GA searches, we used pseudopotentials generated using the following electronic configurations: $3s^2 3p^1 3d^0$ and $2s^2 2p^4$ with cutoff radii of 1.6 and 1.4 a.u. for silicon and oxygen, respectively. They required a cutoff energy of 40 Ry. Brillouin zone integration was performed using the Monkhorst–Pack sampling scheme [40] over k-point meshes of spacing $2\pi \times 0.05$ Å^{-1}. In the structure relaxation steps, constant-pressure variable-cell-shape molecular dynamics (MD) [41,42] was used. Candidate structures obtained with the GA were refined and their static enthalpies and QHA free energies were calculated using harder pseudopotentials, more suitable for the extreme pressures addressed here [29]. The valence electronic configurations of these harder pseudopotentials were $2s^2 2p^6 3s^1 3p^0$ and $2s^2 2p^4 3d^0$ with cutoff radii of 1.2 and 1.0 a.u. for silicon and oxygen, respectively. Their cutoff energy was 400 Ry. We used density functional perturbation theory to compute dynamical matrices at $2 \times 2 \times 2$ q-point mesh for all phases [43,44]. Phonon frequencies were then calculated by interpolation onto q-point meshes fine enough to achieve convergence of QHA free energy within 1 mRy/FU. All first-principles calculations were performed using the Quantum-ESPRESSO software distribution [45], which has been interfaced with the GA scheme in a fully parallel manner.

3.7.1.2 Results and Discussion

The present GA scheme is quite efficient in this system at high pressure. It takes approximately 10 generations at most to reach ground-state configurations for inspected cases. For instance, at 0.5 TPa the GA search using 4-FU SiO_2 produces a pyrite ground state in just a few generations (as shown in Figure 3.8). At 2 TPa, for 4-FU SiO_2, the cotunnite structure is predicted to be the ground state. For 3-FU SiO_2 at both 0.5 and 2 TPa, the Fe_2P structure is rapidly predicted. For 6-FU and 8-FU SiO_2, Fe_2P and cotunnite structures are predicted, respectively. These results indicate that Fe_2P and cotunnite phases are indeed good candidates for post-pyrite phases of SiO_2.

Fe_2P and cotunnite structures (Figure 3.9) are closely related [46,47]. Both have tricapped triangular prisms as structural units with silicon coordination number equal to 9. The lower-pressure phases of SiO_2 (rutile, $CaCl_2$, α-PbO_2, and pyrite) consist of Si octahedra. In the α-PbO_2 structure, shifts of silicons from the octahedral center to the middle of an octahedral face give rise to tricapped triangular prisms [47]. If all silicons shift in the same direction, then the α-PbO_2 structure transforms into the cotunnite structure. If half of silicons shift in the opposite direction, then the α-PbO_2 structure changes into the Fe_2P structure. Several other structures with tricapped triangular prisms can be produced by different shifting patterns. Among them, the NbCoB

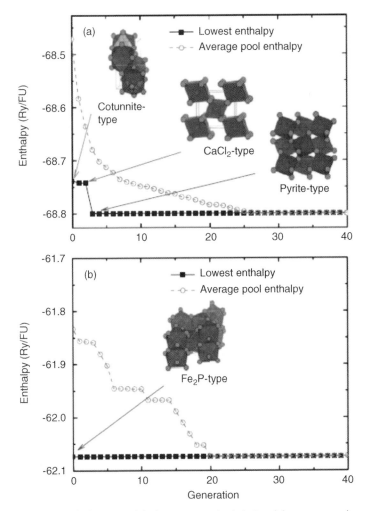

Figure 3.8 The histories of the lowest energy (enthalpy) and the average pool energy (enthalpy) of (a) 4-FU SiO_2 at 0.5 TPa and (b) 3-FU SiO_2 at 2.0 TPa by generation. (From Ref. [67], with permission from American Physical Society.)

structure, whose unit cell consists of 10 FUs [48], is worth noting because it is intermediate between Fe_2P and cotunnite structures. In this structure, Fe_2P and cotunnite structures appear in an alternating pattern (Figure 3.9). Therefore, the NbCoB phase is also a potential post-pyrite phase. Calculated structural parameters of these three potential post-pyrite phases of SiO_2 are given in Table 3.1.

Several other phases appeared in GA pools. Among them, the Li_2ZrF_6 structure [49] is also worth noting. This phase consists of silicon octahedra and is closely related to the α-PbO_2 structure [47], in the same way as Fe_2P and cotunnite structures are related. Fe_2P and cotunnite structures with tricapped triangular prisms are structural counterparts of Li_2ZrF_6 and α-PbO_2 structures with octahedra. Fe_2P-type SiO_2

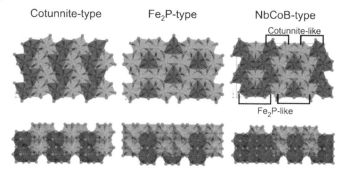

Figure 3.9 Crystal structures of Fe$_2$P-, cotunnite-, and NbCoB-type phases. Blue and light blue spheres denote silicon atoms at different heights. Red small spheres denote oxygen atoms. (From Ref. [67], with permission from American Physical Society.)

Table 3.1 Structural parameters of Fe$_2$P-, cotunnite-, and NbCoB-type SiO$_2$ at 0.8 TPa.

		Fe$_2$P-type SiO$_2$	
	Space group		$P\bar{6}2m$
	(a, c)		(4.120 Å, 2.222 Å)
Si$_1$		2c	(1/3, 2/3, 0)
Si$_2$		1b	(0, 0, 1/2)
O$_1$		3f	(0.2657, 0, 0)
O$_2$		3g	(0.5903, 0, 1/2)
	(B, B')		(2.76 TPa, 2.71)
		Cotunnite-type SiO$_2$	
	Space group		$Pnma$
	(a, b, c)		(4.108 Å, 2.191 Å, 4.853 Å)
Si		4c	(0.2335, 1/4, 0.1357)
O$_1$		4c	(0.3472, 1/4, 0.4347)
O$_2$		4c	(0.9845, 1/4, 0.6670)
	(B, B')		(2.75 TPa, 2.71)
		NbCoB-type SiO$_2$	
	Space group		$Pmmn$
	(a, b, c)		(2.218 Å, 12.010 Å, 4.094 Å)
Si$_1$		4e	(1/4, 0.4470, 0.7612)
Si$_2$		4e	(1/4, 0.3500, 0.2629)
Si$_3$		2b	(1/4, 3/4, 0.2375)
O$_1$		4e	(1/4, 0.5675, 0.5202)
O$_2$		4e	(1/4, 0.6287, 0.025)
O$_3$		2b	(1/4, 3/4, 0.6454)
O$_4$		4e	(1/4, 0.4722, 0.1542)
O$_5$		4e	(1/4, 0.3270, 0.6330)
O$_6$		2a	(1/4, 3/4, 0.0198)
	(B, B')		(2.75 TPa, 2.71)

Source: From Ref. [67], with permission from American Physical Society.
Bulk modulus (B) and its pressure derivative (B') at 0.8 TPa were obtained by the third-order Birch–Murnaghan equation of state.

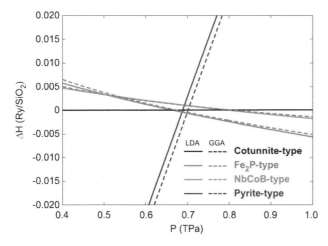

Figure 3.10 Enthalpies of pyrite-, Fe$_2$P-, and NbCoB-type SiO$_2$ with respect to cotunnite-type SiO$_2$. (From Ref. [67], with permission from American Physical Society.)

can be obtained from Li$_2$ZrF$_6$ type by shifting all silicons in the same direction from octahedral centers to octahedral faces. Similarly to the NbCoB structure, there might be an intermediate phase between α-PbO$_2$ and Li$_2$ZrF$_6$ as well. At 2 TPa, most phases in GA pools, including the baddeleyite phase, are found to consist of capped triangular prisms. Enthalpy calculations with harder pseudopotentials confirm that these phases are metastable over the entire pressure range investigated here; Li$_2$ZrF$_6$-type SiO$_2$ has higher enthalpy than α-PbO$_2$ type at all pressures.

Figure 3.10 shows relative enthalpies of several phases of SiO$_2$. The PBE-type generalized gradient approximation (GGA) gives rise to the same results as LDA essentially. Calculated transition pressures by PBE are higher by just ∼10 GPa, as usually expected. Static calculations show that pyrite-type SiO$_2$ transforms to Fe$_2$P type at 0.69 TPa, being consistent with Ref. [34]. This transition pressure is almost identical to the metastable transition pressure between pyrite-type and cotunnite-type SiO$_2$. Although cotunnite and NbCoB phases are metastable over all pressures in static calculations, their enthalpies are very competitive. Below 0.64 TPa, the cotunnite phase has lower enthalpy than Fe$_2$P. Above 0.78 TPa, the NbCoB phase has intermediate enthalpy between those of Fe$_2$P and cotunnite phases. At 1 TPa, enthalpy differences between cotunnite and Fe$_2$P and between NbCoB and Fe$_2$P phases are just 0.006 and 0.004 Ry/FU, respectively. At 2 TPa, these differences increase to 0.017 and 0.01 Ry/FU, respectively (at most ∼3000 K). Phonon calculations show that all three phases are dynamically stable beyond ∼0.4 TPa.

Figure 3.11 a shows the phase diagram of SiO$_2$ predicted by the QHA. Post-pyrite transitions to Fe$_2$P and cotunnite phases have negative Clapeyron slopes. This results from the increase in CNs and bond lengths in Fe$_2$P and cotunnite phases, which increases density of states of low-frequency vibrations and vibrational entropies across these post-pyrite transitions [52]. In contrast, the phase boundary between Fe$_2$P and cotunnite phases has a normal positive Clapeyron slope. Below

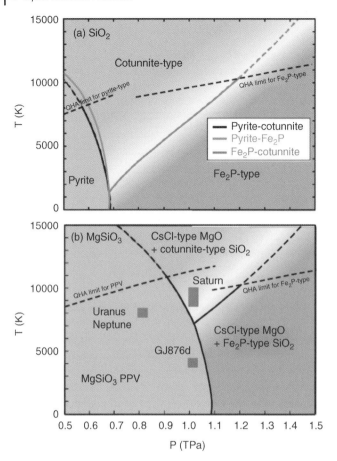

Figure 3.11 Pressure–temperature phase diagram of (a) SiO_2 and (b) dissociation of $MgSiO_3$ PPV into MgO and SiO_2. Free energy of $MgSiO_3$ PPV published in Ref. [7] is used. The transformation between Fe_2P- and cotunnite-type SiO_2 is expected to be gradual. Red areas denote estimated pressure–temperature conditions at core–envelope boundaries in the solar giants and in the GJ876d (see the text). Dashed lines indicate the limit of validity of the QHA. (From Ref. [67], with permission from American Physical Society.)

(above) ∼1500 K, pyrite-type SiO_2 should transform to a Fe_2P-type (cotunnite-type) SiO_2. In contrast, the NbCoB phase does not have a stability field. However, entropic stabilization of disordered structural motifs intermediate between Fe_2P and cotunnite is very likely at high temperatures. Therefore, the Fe_2P-to-cotunnite transition might not be sharp but a rather gradual transformation. Actually, the possibility that a fully disordered, mixed, or even dynamically disordered phase is stable at high temperatures cannot be discarded. However, the cotunnite phase is very stable without phonon instabilities in the pressure and temperature ranges of the phase diagram we presented, suggesting a crossover.

The presence of the Fe_2P phase introduces an additional phase boundary in the dissociation phase diagram of $MgSiO_3$ PPV, as shown in Figure 3.11b. In the icy giants, Uranus and Neptune, the dissociation into MgO and SiO_2 should not occur. In the gas giants, Saturn and Jupiter, the dissociation into CsCl-type MgO and cotunnite-type SiO_2 occurs first. At higher pressures, depending on the internal temperature profiles in these planets, cotunnite-type SiO_2 might transform to Fe_2P type. In GJ876d, a terrestrial exoplanet [53] with ~7.5 Earth masses, conditions estimated at the core–mantle boundary [54] are close to those at the dissociation phase boundary.

Finally, the predicted structural crossover between Fe_2P and cotunnite phases should be fundamental to understanding the high-pressure and high-temperature behavior of AX_2-type compounds. Although pressures for the predicted phenomenon in SiO_2 are challenging to experiments, low-pressure analogues could be investigated to validate our predictions. MgF_2 is particularly suitable because it has a very similar sequence of phase transitions to SiO_2. In combination with NaF, it forms $NaMgF_3$ perovskite, which has the same sequence of predicted phase transitions as $MgSiO_3$ perovskite [55] including dissociation into elementary fluorides/oxides. There are still unresolved questions in the experimental high-pressure behavior of $NaMgF_3$ and MgF_2 [56]. The structural crossover between Fe_2P and cotunnite phases might be part of the answer.

3.7.2
Ultrahigh-Pressure Phases of Ice

H_2O ice is one of the most abundant planet-forming materials and its phase diagram is of fundamental scientific interest. It is crucial for modeling the interiors of icy solar giants (Uranus and Neptune), icy satellites, and also unique ocean planets being discovered presently. Up to now, 16 crystalline phases have been identified experimentally [58–61]. Ice X is the highest-pressure phase among those identified experimentally. In this ionic phase, which exists above ~70 GPa, oxygen atoms form a bcc lattice and hydrogen atoms are located at the midpoint between two nearest-neighbor oxygen atoms. However, in Uranus and Neptune, pressure at their core–envelope boundary is estimated to be as high as 0.8 TPa [28]. Very recently, Neptune-sized icy exoplanets have been discovered [62,63]. Transitions beyond ice X can occur under these high-pressure conditions, and they are essential for modeling the interiors of these planets. Previously, Benoit *et al.* [64] and Caracas [65] predicted a phase with *Pbcm* symmetry as the next high-pressure phase after ice X (up to 0.3 TPa). More recently, Militzer and Wilson proposed transitions from the *Pbcm* to *Pbca* and *Cmcm* phases at 0.76 and 1.55 TPa, respectively [66]. Interestingly, they predicted the *Cmcm* phase to be metallic, an important property for understanding the origin of magnetic fields in the giants. While these studies surely predict the existence of unique phases, their searches using molecular dynamics and phonon instability calculations explored limited regions of phase space producing a structure relatively close to that of ice X. In this section, we report structures of solid H_2O in the TPa pressure regime. They were found using a GA in which a set of empirical potential parameters are continuously adapted using DFT, as described in Section 3.3.

3.7.2.1 Computational Details

The search for the lowest-enthalpy structures of ultrahigh-pressure ice is based on the Deaven–Ho GA scheme [1]. This method, combined with first-principles calculations, has been proven to work very efficiently [3,4,36]. However, first-principles calculations are computationally very demanding for the GA scheme when there are a large number of atoms. On the other hand, GA searches using empirical model potentials are fast but suffer from inaccuracies that can lead the search to wrong structures. However, most of the child structures generated in GA are actually not favorable except in the simplest problems; many false structures have to be tried before hitting on the correct structure. Most of the computer time is spent in relaxing a large number of false candidate structures. For a typical 20 atoms/cell system, ~600 local optimizations are required to reach the global minimum [68]. Each optimization normally consists of 50–100 DFT ionic steps with a typical conjugate gradient method. For larger systems, the number of local optimizations and DFT step time will increase exponentially and become untractable with current computer power. To remedy this, we introduced a structure search technique, the adaptive GA. In the scheme, we employ auxiliary model potentials to estimate the energy ordering of different competing geometries in a preliminary stage. Parameters of the auxiliary potentials are adaptively adjusted to reproduce first-principles results during the course of the GA search. This adaptive approach also allows the system to hop from one basin to another in the energy landscape, leading to efficient sampling of configuration space.

In our study, we found that the packing geometry, volumes (pressures), and energy ordering of ice crystal phases at high pressures (>0.5 TPa) can be described relatively well by simple Lennard-Jones potentials:

$$V_{\text{LJ}} = 4\varepsilon\left[\left(\frac{\sigma}{r}\right)^{12} - \left(\frac{\sigma}{r}\right)^{6}\right]. \tag{3.15}$$

In the H_2O system, a total of six parameters, representing the O–O, O–H, and H–H interactions, are adaptively adjusted to explore structural phase space during the GA search. We initiate the search with random structures and carry out DFT calculations to get their energies, atomic forces, and stress tensor. The LJ potential parameters are fitted to these structures by the force-matching method [69]. Based on this auxiliary potential, we perform a full GA search to yield structural candidates. DFT calculations are carried out on the final GA pool and potential parameters are adaptively adjusted again. These procedures can be done iteratively until the potential parameters converge. The final LJ potential pool population will then be examined with full DFT relaxation. Our test shows that using this adaptive scheme less than 200 single-point DFT calculations are required to obtain a converged auxiliary potential. The total number of candidates in the final full DFT relaxation is also limited, usually less than 100 even for large systems. The adaptive GA is ~100 times faster than full DFT GA for medium-sized systems

(20–40 atoms/cell) and much faster in larger systems. Thus, this adaptive scheme allows a very efficient sampling of crystal structures, and we can easily search relatively large systems up to 12 H_2O formula units.

For the auxiliary potential optimization, we use the large-scale atomic/molecular massively parallel simulator (LAMMPS) code [70] and the conjugate gradient method. First-principles calculations are performed using DFT within the Perdew–Burke–Ernzerhof (PBE)-type generalized gradient approximation [50]. Vanderbilt-type pseudopotentials [39] were generated using the following valence electron configurations: $1s^1$ and $2s^2 2p^4$ with cutoff radii of 0.5 and 1.4 a.u. for hydrogen and oxygen, respectively. Candidate structures obtained with the adaptive GA were then refined using a harder oxygen pseudopotential with a valence electron configuration of $2s^2 2p^4 3d^0$ and a cutoff radius of 1.0 a.u. The cutoff energies for plane-wave expansions are 40 and 120 Ry for the softer and the harder pseudopotentials, respectively. Brillouin zone integration was performed using the Monkhorst–Pack sampling scheme [40] over k-point meshes of spacing $2\pi \times 0.5\ \text{Å}^{-1}$. Structural relaxations were performed using variable-cell-shape molecular dynamics. To test for structural stasbility, phonon and vibrational density of states (VDOS) calculations were carried out for candidate structures using density functional perturbation theory [43,44]. Zero-point motion (ZPM) effects are taken into account within the quasiharmonic approximation [35]. All first-principles calculations were performed using the Quantum-ESPRESSO package [45], which has been interfaced with the GA scheme in a fully parallel manner.

3.7.2.2 Results and Discussion

Figure 3.12 shows average values of the pressures and enthalpies calculated by first principles for structures in the adaptive GA pool. The target pressure of this adaptive GA search is 2 TPa. Our results show a fast convergence of the adaptive GA. After

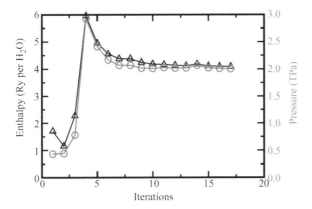

Figure 3.12 Convergence of average DFT pressure (circles) and enthalpy (triangles) of the structures in the LJ potential GA pool as a function of number of adaptive iterations of the potential. The DFT pressure converges to the target pressure of 2 TPa after 10 iterations. The unit cell contains eight H_2O units. (From Ref. [57], with permission from American Physical Society.)

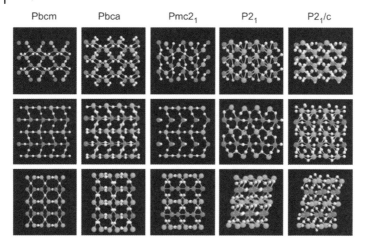

Figure 3.13 Crystal structures of ultrahigh-pressure phases of ice. Blue and red large spheres show oxygen atoms; white small spheres show hydrogen atoms. (From Ref. [57], with permission from American Physical Society.)

10 iterations, LJ potential pressures for structures in the GA pool are almost identical to those obtained from first-principles DFT results. The adaptive GA search successfully predicts three structures: $Pmc2_1$ at 1 TPa, $P2_1$ at 2 TPa, and $P2_1/c$ at 3 TPa (Figure 3.13). They consist of four, four, and eight formula units, respectively. Their structural parameters are listed in Table 3.2. Structures with 3, 5, 6, 7, 9, 10, and 11 formula units were not energetically competitive. *Pbcm* and *Pbca* phases proposed previously were also obtained as metastable phases at these pressures. An examination of the enthalpy differences between the different competing phases (Figure 3.14) shows that the sequence of pressure-induced phase transitions beyond ice X is X → *Pbcm* → *Pbca* → $Pmc2_1$ → $P2_1$ → $P2_1/c$. Static transition pressures are 0.28, 0.75, 0.89, 1.28, and 2.68 TPa, respectively. Zero-point motion strongly affects transition pressures. For X–*Pbcm*, *Pbcm*–*Pbca*, and *Pbca*–$Pmc2_1$ transitions, ZPM increases transition pressures to 0.29, 0.77, and 0.92 TPa, respectively. On the other hand, for $Pmc2_1$–$P2_1$ and $P2_1$–$P2_1/c$ transitions, ZPM greatly decreases transition pressures to 1.14 and 1.96 TPa, respectively.

The structures of these competing phases are closely related. Under compression, ice X transforms to the *Pbcm* phase and then to the *Pbca* phase by means of soft phonon-related deformations [65,66]. In these three phases, all hydrogen atoms are located at the midpoint between two neighboring oxygen atoms. However, during the transition to the $Pmc2_1$ phase, two of the four hydrogen atoms bonded to the O_1 oxygen jump to the octahedral interstitial sites next to two second-nearest-neighbor oxygen atoms. The arrangements of hydrogen atoms are *Pbca*-like around O_1 and *Cmcm*-like around O_2. Therefore, this phase is structurally intermediate between the *Pbca* and *Cmcm* phases. It becomes metastable with respect to the *Cmcm* phase at ∼2.5 TPa. In X, *Pbcm*, *Pbca*, and $Pmc2_1$ phases, OH_4 tetrahedra form the basic structural unit, with connectivity varying in each phase. Across the *Pbca*-to-$Pmc2_1$ transition, two interpenetrating tetrahedral networks transform into a single

Table 3.2 Structural parameters of $Pmc2_1$-, $P2_1$-, and $P2_1/c$-type H_2O.

		$Pmc2_1$-type H_2O at 1 TPa	
	(a, b, c)		(3.087 Å, 1.890 Å, 3.296 Å)
H_1		4c	(0.25235, 0.39936, 0.34531)
H_2		2a	(0, 0.72441, 0.58937)
H_3		2b	(0.5, 0.02194, 0.05554)
O_1		2a	(0, 0.79553, −0.00013)
O_2		2b	(0.5, 0.71235, 0.29490)
		$P2_1$-type H_2O at 2 TPa	
	(a, b, c, β)		(1.711 Å, 3.066 Å, 2.825 Å, 99.83°)
H_1		2a	(0.03146, −0.004190, 0.97579)
H_2		2a	(0.17734, 0.60082, 0.33256)
H_3		2a	(0.25493, 0.38266, 0.73104)
H_4		2a	(0.56268, 0.74399, 0.68952)
O_1		2a	(0.82524, 0.52141, 0.52159)
O_2		2a	(0.34462, 0.75497, 0.01421)
		$P2_1/c$-type H_2O at 3 TPa	
	(a, b, c, β)		(2.921 Å, 2.890 Å, 3.338 Å, 117.86°)
H_1		4e	(−0.49599, 0.19043, 0.44690)
H_2		4e	(−0.14264, 0.13088, −0.07153)
H_3		4e	(−0.24986, −0.50964, −0.27537)
H_4		4e	(0.21275, 0.37363, −0.03277)
O_1		4e	(0.06825, −0.36013, −0.15960)
O_2		4e	(−0.42335, −0.37490, 0.33745)

Source: From Ref. [57], with permission from American Physical Society.

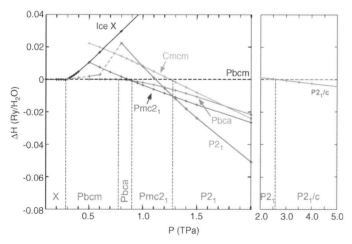

Figure 3.14 Enthalpies of ultrahigh-pressure phases of ice. Dashed vertical lines denote static transition pressures. (From Ref. [57], with permission from American Physical Society.)

network. Locally, hydrogen atoms around O_2 try to keep two interpenetrating networks, while those around O_1 atoms connect two networks into a single one. Tetrahedra around the O_1 atoms are severely distorted compared to those around the O_2 atoms.

The $Pmc2_1$ phase is dynamically stable up to 1.3 TPa. At ~1.5 TPa, a zone-center soft mode appears. This soft mode induces a monoclinic distortion, giving rise to the $P2_1$ phase. The $P2_1$ space group is a subgroup of $Pmc2_1$. In the $P2_1$ phase, the structural unit is no longer OH_4. The coordination number of oxygen atoms increases from 4 to 5. In X, $Pbcm$, $Pbca$, and $Pmc2_1$ phases, phonon dispersions are divided into two groups corresponding to nearly rigid-body and internal OH stretching motions of OH_4 tetrahedra, while this distinction is blurred in the $P2_1$ phase. As a result of a higher coordination number, there is an increase in the H—O bond length and a 2.0% volume reduction across the $Pmc2_1$–$P2_1$ transition. Both $P2_1$ and $P2_1/c$ phases are dynamically stable at least up to 7 TPa, with no soft modes developing under compression.

In contrast with the metallic $Cmcm$ phase predicted by Militzer and Wilson [66], all three phases ($Pmc2_1$, $P2_1$, and $P2_1/c$) have substantial DFT band gaps. In the $P2_1$ and $P2_1/c$ phases, the band gap decreases slowly under compression (Figure 3.15), closing at ~7 TPa in the $P2_1/c$ phase. Although all crystalline phases are insulating, H_2O could be a good conductor at relevant conditions for two reasons: (1) Protons are highly mobile in the oxygen sublattice, producing superionic phases at ultrahigh pressures and temperatures typical of the interiors of icy solar giants and exoplanets; and (2) depending on the temperature, thermally excited carriers also contribute to increase the conductivity [72,79]. So far, only the bcc oxygen sublattice has been considered in the investigation of superionic phases. Our proposed crystal

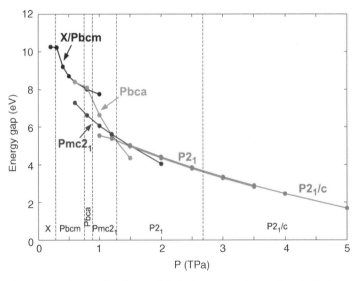

Figure 3.15 Electronic band gaps of ultrahigh-pressure phases of ice. Dashed vertical lines denote static transition pressures. (From Ref. [57], with permission from American Physical Society.)

structures provide the starting points for further investigation of conducting superionic states of these phases at ultrahigh pressures and temperatures. They indicate that the oxygen sublattice should prefer to have hexagonal-derived structures beyond ∼0.4 TPa. This possibly implies a bcc–hcp transition in the superionic phase. Our predictions are consistent with the recent preprint by McMahon [80]. In this work, the searches were carried out by the *ab initio* random structure searching (AIRSS) method but were limited to the 4 H_2O unit cell. Therefore, the $P2_1/c$ phase with 8 H_2O could not be identified. Our global searches up to 12 H_2O provide a more complete phase diagram of ultrahigh-pressure ice.

3.7.3
Structure and Magnetic Properties of Fe–Co Alloys

$Fe_{1-x}Co_x$ alloys have been studied intensively for long time [81–90] because of their high potential of applications in magnetism. $Fe_{1-x}Co_x$ alloys have unique properties in a wide range of Fe concentration: very high saturation magnetization, high critical temperature, and low magnetic anisotropy. These properties are ideal for soft magnetic materials or high field requiring materials [92]. Many experiments have been done to investigate the atomic structure and magnetic properties of $Fe_{1-x}Co_x$ alloys [81–83]. Theoretical studies of the structure and magnetic properties of both disordered and ordered bcc phases have been reported [84–90]. For the disordered bcc phase, magnetic properties of these alloys were investigated by either virtual crystal approximation [84] or coherent potential approximation [85,86]. For the ordered phase, Drautz *et al.* [87,88] used the cluster expansion method to find ground-state structures of Fe-rich alloys and investigated magnetic properties by first-principles calculations. In both disordered and ordered phases, the Fe atom shows an increase in magnetic moment with Co concentration, while the magnetic moment of Co remains almost constant.

In this section, we present the results of a *from-scratch* GA crystal structure search for low-energy crystal structures and the magnetic properties of Fe–Co alloys by first-principles calculations. We will show that there are more stable structures of off-stoichiometric Fe–Co alloys in addition to the well-known B2 (CsCl) structure and that these alloys are highly configurationally degenerate. The magnetic moment of Fe atom increases with Co concentration, while that of Co remains almost constant. The origin of this behavior will be explained.

3.7.3.1 Computational Details
We perform GA crystal structure search for Fe-rich alloys with a combination of classical potential calculation and density functional theory calculations. A small supercell will limit the number of compositions and the number of possible crystal structures of Fe–Co alloys, while the possibility of performing the search with a very big supercell is limited to the computational resources. Previous studies [87–90] showed that a 16-atom supercell is probably big enough for Fe–Co alloy systems to find low-energy crystal structures. Thus, we performed the search with a supercell of 16 atoms, corresponding to eight different compositions of Fe-rich alloys:

Fe_nCo_{16-n}, with n ranging from 8 to 15. Atomic structure relaxations and magnetic moment calculations of Fe–Co alloys were performed using spin-polarized DFT within generalized gradient approximation [50] as implemented in the VASP code [15] with plane-wave basis. We used the projector-augmented wave (PAW) method [93] and the kinetic energy cutoff was 350 eV. In order to minimize the error of total energy calculations, we used the same k-point grid for energy integrations. This grid is equivalent to $16 \times 16 \times 16$ k-points for FeCo B2 (CsCl) unit cell. The formation energy of the alloy is calculated as

$$E_f(Fe_nCo_m) = E(Fe_nCo_m) - n\mu(Fe) - m\mu(Co), \quad (3.16)$$

where $E(Fe_nCo_m)$ is the total energy of Fe_nCo_m alloy and $\mu(Fe)$ and $\mu(Co)$, respectively, are the energy per atom of Fe and Co in the reference structures, which are pure bcc Fe and pure hcp Co metals.

3.7.3.2 Results and Discussion

For each composition of Fe–Co alloys, the GA search for the low-energy crystal structure was performed several times with different starting random configurations. The selected low-energy crystal structures were then fully relaxed by DFT calculations [15] to obtain the final structures. All low-energy crystal structures obtained in the GA search are in bcc lattice with different arrangements of Fe and Co on the lattice. The average volume per atom is 11.5 Å^3/atom, which is consistent with experiment [81] and previous theoretical studies [88]. We also recovered all ground-state crystal structures suggested in the literature [87–90] and found some more low-energy structures in our from-scratch search. These results are interesting because we can search for ground-state structures of an alloy with the only input information of the composition of the alloy. We do not need any extra information from experiment such as the lattice type or lattice constant. All initial atomic coordinates and lattice vectors are generated randomly. They evolved from generation to generation without any constraints. These are the advances of genetic algorithm in comparison with other methods such as cluster expansion [87,88] or direct trial [90], which are limited to the chosen supercells. These advances could explain why we found more low-energy structures in addition to those in the literature, as we will see below.

Except for Fe_8Co_8, where the next lowest-energy structure to the B2 structure is 9.5 meV/atom higher, in all other compositions we found many structures having relative formation energy with respect to the lowest-energy structure within 2 meV/atom, including the literature structures. We show in Figure 3.16 the relative formation energy of low-energy structures of one composition ($Fe_{11}Co_5$). The results for other compositions are similar. The formation energies of all low-energy structures are negative, showing that they are stable structures. Most of these low-energy structures are noncubic in contrast to the cubic supercell of the literature structures. The energy window of 2 meV/atom is about the accurate limit of our DFT calculations; therefore, it is irrelevant to say which crystal structure is the true ground-state structure. Practically, these crystal structures are degenerate, meaning that $Fe_{1-x}Co_x$ alloy is highly configurationally degenerate. This result explains the

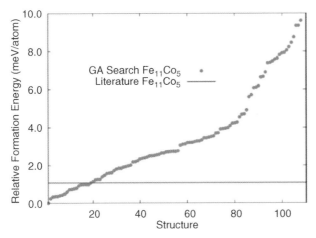

Figure 3.16 The relative formation energies of $Fe_{11}Co_5$ low-energy structures in the unit of meV/atom. (From Ref. [91], with permission from American Institute of Physics.)

difficulty in determining crystal structure of off-stoichiometric $Fe_{1-x}Co_x$ alloys in experiment and could open a new avenue for atomistic manipulation of $Fe_{1-x}Co_x$ alloys by means of changing the conditions of sample preparation process, such as cooling rate, quenching temperature, or external field.

Now we consider the magnetic moments of Fe and Co atoms in $Fe_{1-x}Co_x$ alloys. In Figure 3.17, we plot the average magnetic moments of Fe and Co atoms in various compositions of $Fe_{1-x}Co_x$ alloys. We calculated the average overall structures with relative formation energy smaller than 10 meV/atom. The magnetic moments of Fe and Co at 100 wt% Fe are the magnetic moments of Fe and Co atoms in pure bcc Fe and pure hcp Co metals, respectively. From this figure we see that the magnetic moment of Co is almost constant at $1.74\,\mu_B$ over the whole range of Fe-rich $Fe_{1-x}Co_x$

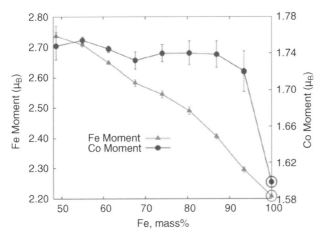

Figure 3.17 Average magnetic moments of Fe and Co in Fe-rich $Fe_{1-x}Co_x$ alloy. (From Ref. [91], with permission from American Institute of Physics.)

alloy, while that of Fe increases with Co concentration. It increases from $2.21\mu_B$ for pure bcc Fe to $2.74\ \mu_B$ in Fe_8Co_8 alloys. This result is consistent with previous theoretical study [88]. In the low Co concentration range ($\leq$20 wt% Co), the increase in Fe magnetic moment is almost linear with a slope of 1.5 μ_B/% from our linear fitting. This slope is quite comparable with the experimental slope of 1.7 μ_B/% [83] for low Co concentration ($\leq$25 wt% Co). The discrepancy between theoretical and experiment results could be attributed to the lack of the orbital contribution to the magnetic moment, as pointed out in previous studies [88,89]. The theoretical values here include only the spin contribution to the moment. From the variations of the magnetic moments of Fe and Co, the average magnetic moment per atom of $Fe_{1-x}Co_x$ alloys has the Slater–Pauling curve with the maximum between 20 and 26 Co wt% (between $Fe_{12}Co_4$ and $Fe_{13}Co_3$).

The increase in Fe magnetic moment with Co concentration implies that Fe magnetic moment would depend directly on the number of Co nearest neighbors as Parette and Mirebeau commented [83]. In order to investigate in more detail and quantitatively the dependence of magnetic moment of Fe atom on the local environment (nearest neighbor only), we calculated the average magnetic moment (overall low-energy structures with relative formation energy smaller than 10 meV/atom) of Fe atoms as a function of the number of Co nearest neighbors. The result of this calculation is shown in Figure 3.18. We see that the magnetic moment of Fe atom increases almost linearly with the number of Co nearest neighbors for all compositions we are considering here. For high Co concentration alloys ($\geq$20 wt% Co), the magnetic moments of Fe atom are almost same for different concentrations except for a little scattering at zero Co nearest neighbors. We can fit only one line with the slope of 0.0452 to data for high Co concentration alloys as plotted in Figure 3.18. This means that the magnetic moment of Fe atom is strongly dependent

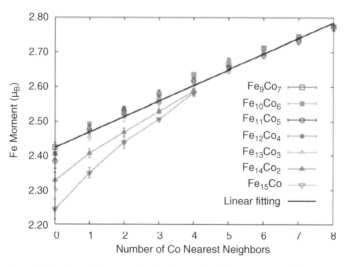

Figure 3.18 Magnetic moment of Fe atom with different number of Co nearest neighbors. (From Ref. [91], with permission from American Institute of Physics.)

on the local environment; especially, the first neighbor atoms dominate the magnetic state of Fe atom. The effect of the outer shell neighbors is quite small and it is smaller for a more number of Co atoms at the first nearest neighbors. This convergent behavior is also observed in low Co concentration alloys. Going to low Co concentration alloys, the magnetic moment of Fe drops suddenly at Co concentration of ∼13 wt%. The drop is more profound for lower Co concentration alloys, indicating that the outer shells here are more effective than those of the high Co concentration alloys. From this strong dependence of magnetic moment of Fe atom on the number of Co nearest neighbors, we can explain the increase in the average magnetic moment of Fe atom with Co concentration. As the Co concentration increases, the average number of Co nearest neighbors increases, inducing an increase in average magnetic moment of Fe atom in the alloy.

In conclusion, we performed a global search for low-energy crystal structures of Fe-rich $Fe_{1-x}Co_x$ alloys by genetic algorithm and first-principles calculations. We found that low-energy structures of $Fe_{1-x}Co_x$ alloys are highly degenerate structures on a bcc lattice. This result could open a new avenue for manipulation of atomic structure of $Fe_{1-x}Co_x$ alloys for various applications. The average magnetic moment of Co in $Fe_{1-x}Co_x$ alloys is almost constant, while that of Fe increases with the concentration of Co, giving the Slater–Pauling curve of the average magnetic moment per atom. The magnetic moment of Fe atoms strongly depends on the number of Co nearest neighbors. The effect of higher-order neighbors is very small for high Co concentration alloys but it is quite profound at small number of Co nearest neighbors in low Co concentration alloys.

References

1 Deaven, D.M. and Ho, K.M. (1995) *Phys. Rev. Lett.*, **75**, 288.
2 Abraham, N.L. and Probert, M.I.J. (2006) *Phys. Rev. B*, **73**, 224104.
3 Oganov, A.R. and Glass, C.W. (2006) *J. Chem. Phys.*, **124**, 244704.
4 Trimarchi, G. and Zunger, A. (2007) *Phys. Rev. B*, **75**, 104113.
5 Trimarchi, G., Freeman, A.J., and Zunger, A. (2009) *Phys. Rev. B*, **80**, 092101.
6 Oganov, A.R. (ed.) (2010) *Modern Methods of Crystal Structure Prediction*, Wiley-VCH Verlag GmbH, Weinheim.
7 Ji, M., Umemoto, K., Wang, C.Z., Ho, K.M., and Wentzcovitch, R. (2010) *Phys. Rev. B*, **84**, 220105(R).
8 Oganov, A.R. and Glass, C.W. (2008) *J. Phys.: Condens. Matter*, **20**, 064210.
9 Oganov, A.R., Lyakhov, A.O., and Valle, M. (2010) *Acc. Chem. Res.*, **44**, 227.
10 Oganov, A.R., Chen, J., Gatti, C., Ma, Y.Z., Ma, Y.M., Glass, C.W., Liu, Z., Yu, T., Kurakevyth, O.O., and Solozhenko, V.L. (2009) *Nature*, **457**, 863.
11 Ma, Y.M., Eremets, M., Oganov, A.R., Xie, Y., Trojan, I., Medvedev, S., Lyakhov, A.O., Valle, M., and Prakapenka, V. (2009) *Nature*, **458**, 182.
12 Oganov, A.R., Glass, C.W., and Ono, S. (2006) *Earth Planet. Sci. Lett.*, **241**, 95.
13 Li, Q., Ma, Y.M., Oganov, A.R., Wang, H.B., Wang, H., Xu, Y., Cui, T., Mao, H.K., and Zou, G.T. (2009) *Phys. Rev. Lett.*, **102**, 175506.
14 Gao, G.Y., Oganov, A.R., Bergara, A., Martinez-Canales, M., Cui, T., Iitaka, T., Ma, Y.M., and Zou, G.T. (2008) *Phys. Rev. Lett.*, **101**, 107002.
15 Kresse, G. and Furthmuller, J. (1996) *Phys. Rev. B*, **54**, 11169.

16 Stillinger, F.H. and Weber, T.A. (1985) *Phys. Rev. B*, **31**, 5262.
17 Tersoff, J. (1988) *Phys. Rev. B*, **38**, 9902; **37**, 6991 (1988).
18 Lenosky, T.J., Sadigh, B., Alonso, E., Bulatov, V.V., Diaz de la Rubia, T., Kim, J., Voter, A.F., and Kress, J.D. (2000) *Model. Simulat. Mater. Sci. Eng.*, **8**, 825.
19 Lennard-Jones, J.E. (1924) *Proc. R. Soc. Lond. A*, **106** (738), 463.
20 Daw, M.S. and Baskes, M. (1984) *Phys. Rev. B*, **29**, 6443.
21 Oganov, A.R. and Valle, M. (2009) *J. Chem. Phys.*, **130**, 104504.
22 d'Avezac, M. and Zunger, A. (2008) *Phys. Rev. B*, **78**, 064102.
23 Murakami, M., Hirose, K., Kawamura, K., Sata, N., and Ohishi, Y. (2004) *Science*, **304**, 855.
24 Oganov, A.R. and Ono, S. (2004) *Nature*, **430**, 445.
25 Tsuchiya, T., Tsuchiya, J., Umemoto, K., and Wentzcovitch, R.M. (2004) *Earth Planet. Sci. Lett.*, **224**, 241.
26 van den Berg, A.P., Yuen, D.A., Beebe, G.L., and Christiansen, M.D. (2010) *Phys. Earth Planet. Interiors*, **178**, 136.
27 Sasselov, D.D. and Valencia, D. (2010) *Sci. Am.*, **303** (2), 38.
28 Guillot, T. (2004) *Phys. Today*, **57** (4), 63.
29 Umemoto, K., Wentzcovitch, R.M., and Allen, P.B. (2006) *Science*, **311**, 983.
30 Kuwayama, Y., Hirose, K., Sata, N., and Ohishi, Y. (2005) *Science*, **309**, 923.
31 Haines, J., Leger, J.M., Gorelli, F., Klug, D.D., Tse, J.S., and Li, Z.Q. (2001) *Phys. Rev. B*, **64**, 134110.
32 Grocholski, B., Shim, S.-H., and Prakapenka, V.B. (2010) *Geophys. Res. Lett.*, **37**, L14204.
33 Oganov, A.R., Gillan, M.J., and Price, G.D. (2005) *Phys. Rev. B*, **71**, 064104.
34 Tsuchiya, T., Tsuchiya, J., Metsue, A., and Ishikawa, T. (2010) *Acta Mineral. Petrogr.*, **6**, 810.
35 Wallace, D. (1972) *Thermodynamics of Crystals*, John Wiley & Sons, Inc., New York.
36 Ji, M., Wang, C.-Z., and Ho, K.-M. (2010) *Phys. Chem. Chem. Phys.*, **12**, 11617.
37 Ceperley, D.M. and Alder, B.J. (1980) *Phys. Rev. Lett.*, **45**, 566.
38 Perdew, J.P. and Zunger, A. (1981) *Phys. Rev. B*, **23**, 5048.
39 Vanderbilt, D. (1990) *Phys. Rev. B*, **41**, R7892.
40 Monkhorst, H.J. and Pack, J.D. (1976) *Phys. Rev. B*, **13**, 5188.
41 Wentzcovitch, R.M. (1991) *Phys. Rev. B*, **44**, 2358.
42 Wentzcovitch, R.M., Martins, J.L., and Price, G.D. (1993) *Phys. Rev. Lett.*, **70**, 3947.
43 Baroni, S., de Gironcoli, S., Dal Corso, A., and Giannozzi, P. (2001) *Rev. Mod. Phys.*, **73**, 515.
44 Giannozzi, P., de Gironcoli, S., Pavone, P., and Baroni, S. (1991) *Phys. Rev. B*, **43**, 7231.
45 Giannozzi, P. et al. (2009) *J. Phys.: Condens. Matter*, **21**, 395502.
46 Dera, P., Lavina, B., Borkowski, L.A., Prakapenka, V.B., Sutton, S.R., Rivers, M.L., Downs, R.T., Boctor, N.Z., and Prewitt, andC.T. (2008) *Geophys. Res. Lett.*, **35**, L10301.
47 Hyde, B.G. and Andersson, S. (1989) *Inorganic Crystal Structures*, John Wiley & Sons, Inc., New York.
48 Krypyakevich, P.I., Kuz'ma, Y.B., Voroshilov, Y.V., Shoemaker, C.B., and Shoemaker, D.P. (1971) *Acta Crystallogr. Sect. B*, **27**, 257.
49 Brunton, G. (1973) *Acta Crystallogr. Sect. B*, **29**, 2294.
50 Perdew, J.P., Burke, K., and Ernzerhof, M. (1996) *Phys. Rev. Lett.*, **77**, 3865; **78**, 1396.
51 Wentzcovitch, R.M., Yu, Y., and Wu, Z. (2010) *Rev. Mineral. Geochem.*, **71**, 59.
52 Navrotsky, A. (1980) *Geophys. Res. Lett.*, **7**, 709.
53 Rivera, E.J. et al. (2005) *Astrophys. J.*, **634**, 625.
54 Valencia, D., O'Connell, R.J., and Sasselov, D. (2006) *Icarus*, **181**, 545.
55 Umemoto, K., Wentzcovitch, R.M., Weidner, D.J., and Parise, J.B. (2006) *Geophys. Res. Lett.*, **33**, L15304.
56 Martin, C.D., Crichton, W.A., Liu, H., Prakapenka, V., Chen, J., and Parise, J.B. (2006) *Geophys. Res. Lett.*, **33**, L11305.
57 Ji, M., Umemoto, K., Wang, C.Z., Ho, K.M., and Wentzcovitch, R.M. (2011) *Phys. Rev. B*, **84**, 220105.
58 Petrenko, V.F. and Whitworth, R.W. (1999) *Physics of Ice*, Oxford University Press, Oxford, UK.

59 Lobban, C., Finney, J.L., and Kuhs, W.F. (1998) *Nature*, **391**, 268.
60 Salzmann, C.G., Radaelli, P.G., Hallbrucker, A., Mayer, E., and Finney, J.L. (2006) *Science*, **311**, 1758.
61 Salzmann, C.G., Radaelli, P.G., Mayer, E., and Finney, J.L. (2009) *Phys. Rev. Lett.*, **103**, 105701.
62 Butler, P., Vogt, S.S., Marcy, G.W., Fischer, D.A., Wright, J.T., Henry, G.W., Laughlin, G., and Lissauer, J. (2004) *Astrophys. J.*, **617**, 580.
63 Lissauer, J.J. *et al.* (2011) *Nature*, **470**, 53.
64 Benoit, M., Bernasconi, M., Focher, P., and Parrinello, M. (1996) *Phys. Rev. Lett.*, **76**, 2934.
65 Caracas, R. (2008) *Phys. Rev. Lett.*, **101**, 085502.
66 Militzer, B. and Wilson, H.F. (2010) *Phys. Rev. Lett.*, **105**, 195701.
67 Wu, S., Umemoto, K., Ji, M., Wang, C.Z., Ho, K.M., and Wentzcovitch, R.M. (2011) *Phys. Rev. B*, **83**, 184102.
68 Gao, G., Oganov, A.R., Wang, H., Li, P., Ma, Y., Cui, T., and Zou, G. (2010) *J. Chem. Phys.*, **133**, 144508.
69 Brommer, P. and Gähler, F. (2006) *Philos. Mag.*, **86**, 753.
70 Plimpton, S. (1995) *J. Comput. Phys.*, **117**, 1.
71 Giannozzi, P. *et al.* (2009) *J. Phys.: Condens. Matter*, **21**, 395502.
72 Cavazzoni, C., Chiarotti, G.L., Scandolo, S., Tosatti, E., Bernasconi, M., and Parrinello, M. (1999) *Science*, **283**, 44.
73 Chau, R., Mitchell, A.C., Minich, H.W., and Nellis, W.J. (2001) *J. Chem. Phys.*, **114**, 1361.
74 Goldman, N., Fried, L.E., Feng, I., Kuo, W., and Mundy, C.J. (2005) *Phys. Rev. Lett.*, **94**, 217801.
75 Goncharov, A.F., Goldman, N., Fried, L.E., Crowhurst, J.C., Feng, I., Kuo, W., Mundy, C.J., and Zang, J.M. (2005) *Phys. Rev. Lett.*, **94**, 125508.
76 Mattsson, T.R. and Desjarlais, M.P. (2006) *Phys. Rev. Lett.*, **97**, 017801.
77 Schwegler, E., Sharma, M., Gygi, F., and Galli, G. (2008) *Proc. Natl. Acad. Sci. USA*, **105**, 14779.
78 French, M., Mattsson, T.R., Nettelmann, N., and Redmer, R. (2009) *Phys. Rev. B*, **79**, 054107.
79 Redmer, R., Mattsson, T.R., Nettelmann, N., and French, M. (2011) *Icarus*, **211**, 798.
80 McMahon, J.M. (2011) *Phys. Rev. B*, **84**, 220104.
81 Ellis, W.C. and Greiner, E.S. (1941) *Trans. Am. Soc. Met.*, **29**, 415.
82 Bardos, D.I. (1969) *J. Appl. Phys.*, **40**, 1371.
83 Parette, G. and Mirebeau, I. (1989) *Physica B*, **156**, 721.
84 Victora, R.H. and Falicov, L.M. (1984) *Phys. Rev. B*, **30**, 259.
85 Richter, R. and Eschrig, H. (1988) *J. Phys. F: Met. Phys.*, **18**, 1813.
86 Turek, I., Kudrnovsky, J., Drchal, V., and Weinberger, P. (1994) *Phys. Rev. B*, **49**, 3352.
87 Drautz, R., Diaz-Ortiz, A., Fahnle, M., and Dosch, H. (2004) *Phys. Rev. Lett.*, **93**, 067202.
88 Diaz-Ortiz, A., Drautz, R., Fahnle, M., Dosch, H., and Sanchez, J.M. (2006) *Phys. Rev. B*, **73**, 224208.
89 MacLaren, J.M., Schulthess, T.C., Butler, W.H., Sutton, R., and McHenry, M. (1999) *J. Appl. Phys.*, **85**, 4833.
90 Wu, D., Zhang, Q., Liu, J.P., Yuang, D., and Wu, R. (2008) *Appl. Phys. Lett.*, **92**, 052503.
91 Nguyen, M.C., Zhao, X., Ji, M., Wang, C.Z., Harmon, B., and Ho, K.M. (2012) *J. Appl. Phys.*, **111**, 07E338.
92 Sundar, R.S. and Deevi, S.C. (2005) *Int. Mater. Rev.*, **50**, 157.
93 Kresse, G. and Joubert, D. (1999) *Phys. Rev. B*, **59**, 1758.

4
Optimization of Atomic Clusters

Following the pioneering development paper of Deaven and Ho [1], genetic algorithms (GAs) have been applied extensively to determine the structure of atomic clusters of various kinds. The scientific curiosity associated with research on cluster structure stems from the fact that they have properties significantly different from the bulk materials made of the same elements, or same alloys, or same compounds. In the size range of a few to a few million atoms, the atomic structure is also different from that of the bulk material, and this is the reason why the properties are so different. Another reason for the sustained scientific interest in clusters is more subtle, and stems from the possibility of controlling a whole host of properties through the size of the clusters (sometimes also called nanoparticles when their diameters are in the tens of nanometer range). For example, if for a given material we understand clearly how the structure, shape, and properties (electronic, optical, and chemical properties) vary as a function of size, and especially at what size the bulk material behavior kicks in, then we have a great deal of knowledge about the material that can readily enable a wide range of technological applications.

4.1
Alloys, Oxides, and Other Cluster Materials

Several useful reviews on the subject of GA optimization of clusters already exist, in which the GA optimization of clusters is analyzed either specifically, in and of itself [2,3], or as part of larger discussions on global optimization methods [4–7]. As friendly reminder to the reader, in this book the term "genetic algorithm" is used to encompass both the *binary representation* of the atomic structure (this binary representation by itself is often called genotype representation, and the algorithm is often called genetic) and the *real-space representation* (which by itself is often called phenotype and the algorithms are sometimes called evolutionary). Since the binary representation for atomic structure determination adds to the difficulty of the problem by ignoring correlations imposed by the chemistry (bonding) of the systems, we are dealing largely with the real-space representation that involves relaxations to the nearest local minimum before any decision is taken regarding the faith of a newly created structure.

Atomic Structure Prediction of Nanostructures, Clusters and Surfaces,
First Edition. Cristian V. Ciobanu, Cai-Zhuang Wang, and Kai-Ming Ho.
© 2013 Wiley-VCH Verlag GmbH & Co. KGaA. Published 2013 by Wiley-VCH Verlag GmbH & Co. KGaA.

In terms of elemental clusters, we have already described in Chapter 2 group IV elements (carbon, silicon, and germanium) for which the GA procedure has been applied to determine ground-state structures in various size regimes. In addition to group IV elemental clusters, other important examples of clusters that can serve not only as example of routine GA applications, but also as evidence of striking changes in properties and structure as the size of the cluster is varied are

- metallic clusters of Al, Au, Ag, Ni, Be, Zn, V, Pb [8–14], and others;
- bimetallic alloy clusters of Co–Cu, Ni–Al, Co–Rh, Cu–Au, and many others [15–19];
- oxide clusters of TiO_2, SiO_2, and MgO [20–22];
- metal phosphides [23];
- metal fluorides [24];
- passivated silicon clusters [25,26];
- water clusters [27,28].

Various available examples of global minima of metallic alloys are shown in Figure 4.1, and for oxide nanoparticles in Figure 4.2. In the case of alloys, be they metallic or not, it is a fundamental curiosity to figure out how a given composition of the cluster is accommodated: is the minority species segregated at the surface of the surface? uniformly distributed through the cluster? layered in some way? or is there a compound with specific stoichiometry being formed, while the possible excess of one of the components is being segregated? These are very important and intriguing questions that do not have a general answer but depend on the two elements combined

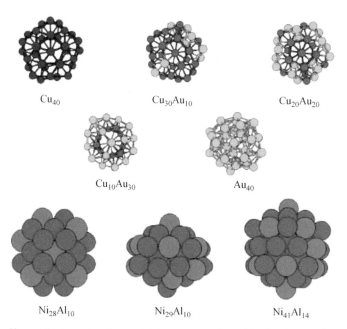

Figure 4.1 Examples of ground-state structures of metallic alloy clusters of Cu–Au and Ni–Al. (From Ref. [2], with permission from Royal Society of Chemistry.)

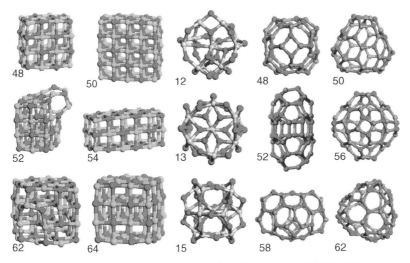

Figure 4.2 Diverse examples of oxide nanoparticles (clusters). (From Ref. [5], with permission from Nature Publishing Group.)

to form the cluster. To add to this mystery, it is not hard to realize that the answer to these questions is not solely determined by the composition and type of elements combined, but it is determined by the size of the cluster as well! For example, in the size regime where an elemental cluster has bulk behavior, to accommodate impurities would mean to create stress in the cluster; in turn, this may mean that the impurity stays at the surface. However, if the cluster is so small that the bulk structure is far from taking effect, the atomic arrangement and bonds differ significantly and an impurity might be able to locate itself inside the cluster as well.

It is also important to mention that GA has been successfully applied to Lennard-Jones clusters as well [29], because for these clusters there was a significant body of work obtained, in particular, via the basin hopping Monte Carlo method [30] (briefly discussed in Chapter 6); such body of work serves as reference for the GA findings concerning the structure of Lennard-Jones clusters. The following sections include full descriptions of two situations that are less frequently encountered in the literature (compared, for example, with the structure of elemental clusters or that of metallic alloy clusters). The former situation is likely to emerge quickly as a frequent necessity when dealing with clusters on surfaces [31] (see Section 4.2), while the latter solves a historically important problem (the Thomson problem of finding the lowest-energy configuration of n point charges on a unit sphere [32]; see Section 4.3). Neither of these are found in the reviews already mentioned [2–7].

4.2
Optimization of Substrate-Supported Clusters via GA

Magic clusters and ordered arrays of magic-sized structures on semiconductor surfaces [33–36] may play an important role in the fabrication of surface nanostructures with uniform size and shape distribution. Formation of Si magic clusters on the Si(111)-7 × 7 surface has been studied by Hwang et al. [33] using scanning tunneling microscopy (STM). These magic clusters are energetically stable entities on the substrate and are able to diffuse as a whole. It has been showed that Si magic clusters on the Si(111)-7 × 7 surface are the fundamental units in mass transport phenomena, step fluctuations, detachment and attachment of Si atoms at step edges, and epitaxial growth [33]. More recently, ordered arrays of nanoclusters formed upon deposition of Al, In, and Ga on the Si(111)-7 × 7 surface have also been observed experimentally using STM [34–36].

While STM studies give valuable information about the overall shape and the size of the magic clusters on the Si(111)-7 × 7 surface, the images from STM cannot provide exact determination of the detailed atomic structure of the magic clusters. A comprehensive understanding of the structural and dynamical properties of the magic clusters on the surfaces would be very important for precise manipulation of nanostructure assembly on semiconductor substrate and for studying the properties of these assembled nanostructures. In this section, we focus on determining the structures of the Si magic clusters on the Si(111)-7 × 7 surface, which can be subsequently used as a first step toward understanding their electronic properties and future applications. Global structural optimization is developed and performed using GA coupled with the environment-dependent Si tight-binding (TB) potential [37]. The low-energy candidates selected from the GA optimizations were further relaxed using first-principles total energy calculations. Simulated STM images were calculated to compare with experimental observations.

The STM studies by Hwang et al. showed that Si magic clusters appear on the faulted half of the Si(111)-7 × 7 unit cell, and are approximately 1.5 Å higher than the Si adatoms [33]. The mirror symmetry of the 7 × 7 reconstruction along the [1$\bar{1}$2] direction is broken due to the presence of the cluster. The empty-state STM image shows six or three protrusions of nearly equal brightness forming a ring-like structure [33]. The spacing between adjacent protrusions is much larger than the length of the Si—Si bond, indicating that the cluster probably contains more than six Si atoms. The size of the magic cluster is estimated to be between 9 and 15 atoms by Hwang et al. Dynamical studies show that these magic clusters diffuse as a unit and are still observable up to a temperature of 500 °C. While the magic clusters diffuse on the surface, the detachment of those clusters leaves the surface intact. This experimental information is very useful for guiding the design of our atomic structure search strategy as described below.

The genetic algorithm has been used successfully in global structural optimization for isolated medium-sized silicon clusters [38], and also to study the reconstructions of high-index silicon surfaces [39,40]. We note that the total number of atoms involved in a Si(111)-7 × 7 substrate is too large for performing such global minimum search using tight-binding potential. Obviously, the total number of the atoms involved in the GA operations needs to be reduced. Since the magic clusters are found to be located on one half of the 7 × 7 unit cell, our GA optimization is performed only on a half unit cell of the 7 × 7 reconstruction together with the dimers at the boundary between the faulted and the unfaulted halves of the 7 × 7 reconstruction as shown in Figure 4.3. We

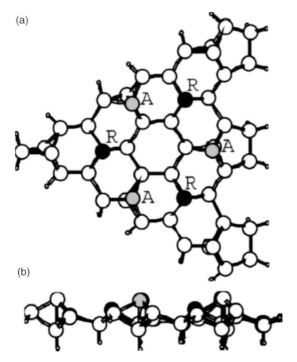

Figure 4.3 The top view (a) and the side view (b) of the half unit cell Si(111)-7 × 7 substrate used in our genetic algorithm. The three adatoms (labeled with A) and the three rest atoms (labeled with R) involved in the GA mating process are shown in gray and black, respectively, whereas the atoms that are not involved in the GA operation are shown in white. Big white circles are the Si atoms, whereas the small circles are the H atoms. The Si dangling bonds at the edges and at the bottom of the unit cell are passivated with hydrogen. (From Ref. [31], with permission from Elsevier.)

also note that the bonding between the magic cluster and the substrate involves mostly the top substrate atoms since diffusion of the magic cluster leaves the surface reconstruction intact. Therefore, only one Si bilayer is used to model the substrate in order to further reduce the number of atoms involved in the GA optimization. The bottom layer of the slab and the Si dangling bonds at the edge of the unit cell are passivated by hydrogen atoms. The final model substrate is shown in Figure 4.3.

Once the substrate is chosen, the procedures of our genetic algorithm are as follows. We generate 10 geometries as initial population by randomly placing a specific number of Si atoms on the half unit cell and then relaxing the structure to a local minimum. The relaxation is done using tight-binding potential with the steepest-descent method. In each subsequent generation, four distinct geometries are randomly picked from the population pool. Then, 12 new geometries are generated from the four parents through the mating operation described below. These children geometries are then relaxed to their nearest local minima. The population of 10 geometries in the pool is updated if any child geometry has different structure and lower energy than any of those in the present population pool. The genetic algorithm was implemented up to 400 generations.

The mating operation produces a child geometry C from the two given parent geometries A and B as follows. For each of the parent geometry, the atoms in the cluster along with three rest atoms and three adatoms (the black and gray atoms in Figure 4.3) on a half unit cell substrate are chosen to be mated. The centers of mass of these chosen atoms in both geometries are then calculated and both parents are cut through the centers of mass with randomly chosen parallel planes. The child geometry C is created by assembling the right half of the chosen atoms in the parent A and the left half of the chosen atoms in the parent B. The resultant atomic configuration of this child C is placed back on the substrate that has three rest atoms and three adatoms removed as illustrated in Figure 4.3. During the mating process, the number of atoms in the resultant child C is required to be the same as that in its parents.

We note that the energy calculations using a one-bilayer substrate cannot differentiate between faulted half and unfaulted half of the Si(111)-7×7 substrate. Therefore, the candidate geometries obtained by our genetic algorithm described above are further optimized using tight-binding molecular dynamics (TBMD) by augmenting the structures with a two-bilayer full 7×7 substrate. The bottom layer of the substrate is passivated by H, and this two-bilayer full 7×7 substrate contains 200 Si atoms and 49 H atoms with a periodic boundary condition parallel to the surface. Two individual optimizations of each structure are done by putting the cluster on either the faulted half or the unfaulted half of the substrate. Finally, the best geometries obtained from genetic algorithm and TBMD optimizations are further studied by first-principles total energy calculations using the same full 7×7 two-bilayer substrate containing 200 Si atoms and 49 H atoms. First-principles calculations in this work have been done using the PWscf package [41], which is based on density functional theory [42] under local density approximations and uses pseudopotential and plane-wave basis. The Ceperley–Alder functional [43] parameterized by Perdew and Zunger [44] is used for the exchange-correlation energy functional. The kinetic energy cutoff is set to be 12 Ry and only the Γ-point is used in the Brillouin zone sum. A vacuum layer of 12 Å is used in the calculations, and the structures are optimized until the forces are smaller than 0.025 eV/Å.

We have performed global structural optimizations for Si clusters on Si(111)-7×7 for sizes ranging from $n = 6$ to $n = 14$. The lowest-energy structure of each size is shown in Figure 4.4, where only the faulted half of the 7×7 unit cell is shown for clarity. In all structures, one of the substrate adatoms is found to be incorporated into the Si cluster. This incorporation will leave three backbond atoms unbonded and the Si cluster will subsequently anchor itself to these backbond atoms. This structural feature justifies the necessity of including the three adatoms and the three rest atoms of the substrate in our GA mating operation. For $n = 6$, the structure obtained by placing six Si atoms at the bright spots as indicated in the STM experiment [33] is not stable according to our first-principles calculations. Instead, our GA search for $n = 6$ found that a six-member ring with an extra Si atom at the center is the lowest-energy structure, as shown in Figure 4.4a. The bonds between one of the substrate adatoms and the corresponding three backbond atoms are broken and the adatom is bonded with the deposited atoms. Another noteworthy feature is that for $n < 12$ the

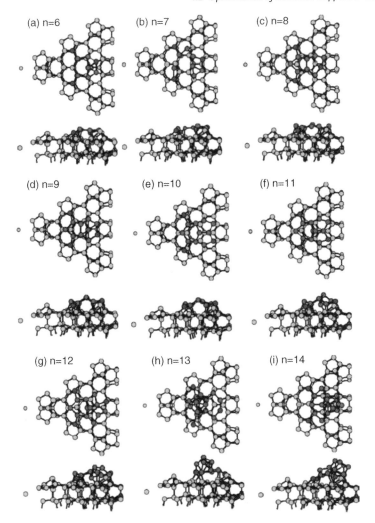

Figure 4.4 The geometries of clusters with sizes ranging from $n = 6$ to 14 obtained by GA. The Si atoms in the cluster are darker, while the three adatoms and the three rest atoms in the substrate that are involved in crossovers are lighter. The geometries shown are optimized with a full 7×7 unit cell. For clarity, only the faulted half of the Si(111)-7×7 surface is shown. (From Ref. [31], with permission from Elsevier.)

cluster exhibits a nearly planar geometry on the surface, whereas for $n \geq 13$ the cluster starts to show a three-dimensional growth.

In order to determine the size n of the magic cluster observed in experiments and to study the relative stability of different sizes of Si clusters on the Si(111) surface, we calculated the surface energy per cluster atom at the surface (E_s) as

$$E_s = \frac{E - E_{\text{sub}} - n\mu}{n}, \tag{4.1}$$

where E is the total energy of the cluster on the Si(111)-7 × 7 substrate, E_{sub} is the energy of the Si(111)-7 × 7 substrate, and μ is the energy of bulk Si (per atom) in crystalline diamond structure. We first checked the surface energy ordering of different-sized clusters on the surface using tight-binding calculations. We found that the energy ordering is the same regardless of the choice of substrate, that is, half unit cell or the full 7 × 7 substrate, and the thickness of the substrate. This result indicates that the reduced Si substrate we used in our GA global search is sufficient for the present study. The plots of surface energies versus the size of the Si cluster obtained from the tight-binding and first-principles calculations are shown in Figure 4.5a and b, respectively. The plots show that Si cluster on the faulted half of the 7 × 7 substrate is energetically more favorable than that on the unfaulted half. This agrees with STM observations that statistically more magic clusters reside on

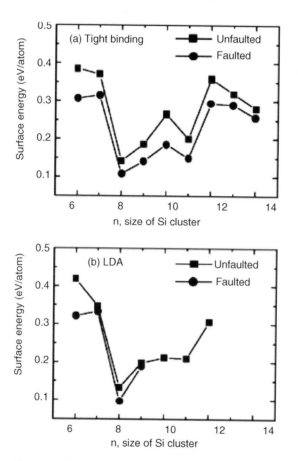

Figure 4.5 The surface energy per cluster atom calculated by (a) tight-binding potential and (b) first-principles calculations. The curve with circles (squares) corresponds to the Si cluster being placed on the faulted (unfaulted) half of the Si(111)-7 × 7 substrate. (From Ref. [31], with permission from Elsevier.)

the faulted half than on the unfaulted half of the Si(111)-7 × 7 surface. We found that there is a sharp dip in the surface energy at $n = 8$, which is the most stable Si cluster on Si(111)-7 × 7 among the sizes that we considered. By comparing Figure 4.5a and b, we found that our tight-binding Si potential can reproduce the overall trend of the surface energies versus size of the Si cluster.

Our calculations suggest that the magic Si cluster on Si(111)-7 × 7 surface should correspond to the $n = 8$ structure, as shown in Figure 4.4c (as well as Figure 4.7). This structure is energetically most favorable when compared with the other structures. The presence of this magic cluster ($n = 8$) minimizes its own dangling bonds while saturating those of the clean Si(111)-7 × 7 reconstruction surface, and does not introduce extra strained bonds or fivefold Si coordination. We calculated simulated STM images using our atomic model of the Si magic cluster and compared them with experimental STM images. The simulated STM images are calculated according to the theory of Tersoff and Hamann [45] by tracing the height profile of constant tunneling current to produce a three-dimensional surface, which is then projected onto the xy-plane to produce our simulated STM images. As a benchmark, the simulated STM image for a clean Si(111)-7 × 7 reconstruction surface is shown in Figure 4.6 in comparison with the experimental image. In addition to the bright spots corresponding to the adatom positions (12 per 7 × 7 unit cell) as seen in the experiment, our simulated STM image also exhibits dimmer spots at the rest atom positions due to the higher resolution in the theoretical calculation. Our stimulated image for the clean 7 × 7 surface is also in a good agreement with other studies [46]. Next, our calculated empty-state images with a bias of +1 V for silicon clusters on Si(111)-7 × 7 surface with the cluster sizes $n = 8$ and 9 are shown in Figure 4.7 together with the experimental STM images reproduced from Ref. 33. Both simulated empty-state images show a ring-like feature. Within the ring, one circular bright spot and two elongated spots can be found for $n = 8$, while there are six bright spots for $n = 9$. Experimental STM images are also shown in Figure 4.7 for comparison. The experimental image on the left

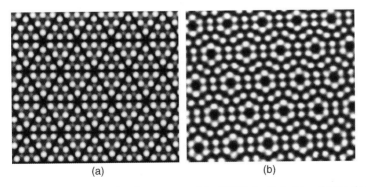

Figure 4.6 (a) The simulated STM image of the Si(111)-7 × 7 at a bias of +1 V. (b) The experimental STM image of the Si(111)-7 × 7 at a bias of +1.5 V. (Courtesy of Miron Hupalo. From Ref. [31], with permission from Elsevier.)

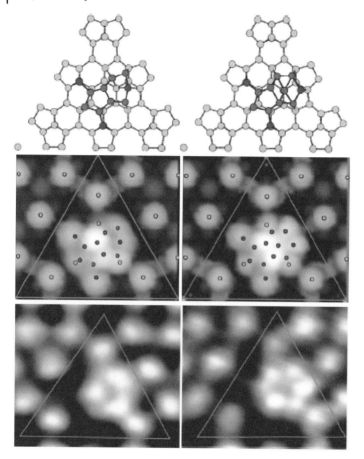

Figure 4.7 *Top*: The positions of the cluster atoms for $n = 8$ (left) and $n = 9$ (right), the adatoms, and rest atoms on the Si(111)-7×7 are indicated and are depicted in the same way as in Figure 4.4. *Center*: The simulated STM images of the $n = 8$ Si cluster (left) and $n = 9$ Si cluster (right) with a bias of $+1$ V. *Bottom*: The experimental STM images of two clusters taken at the sample biases of $+1.8$ V (left) and $+1.2$ V (right) with two different tips are reproduced from Ref. [33]. (From Ref. [31], with permission from Elsevier.)

shows three spots, whereas the other STM image shows six equally bright spots. The difference between the two experimental images was attributed to tip effects. We note that our lowest-energy $n = 8$ structure has actually two degenerate states: one is shown in Figure 4.7 and the other is the mirror image of the first one along the [$\bar{1}12$] direction. We have performed a TBMD simulation for our $n = 8$ structure at elevated temperature and found that there can be flipping between the two degenerate $n = 8$ states. Due to the time averaging of the two states, the experimental STM image would have a total of six spots instead of three as shown in the lower-right corner of Figure 4.7. However, there may be interactions between the STM tip and the magic cluster such that one of the degenerate geometries is pinned, and hence the experiments can also observe three spots instead of six for certain STM tips as shown in the lower-left corner of Figure 4.7. Apart from the ring, both our simulated images show three additional spots due to the three unsaturated atoms of the Si clusters, which are not observed in STM experiments.

In summary, global structural optimization using GA was performed to search for the ground-state structures of Si clusters on the Si(111)-7 × 7 surface for $n = 6$ up to 14. We found that Si clusters prefer to reside on the faulted half of the 7 × 7 substrate for all the sizes studied. When the Si cluster contains 13 atoms or more, it begins to exhibit three-dimensional growth. Placing six Si atoms according to the experimental bright spot positions does not produce a stable structure. Instead, our study suggests that the structure with $n = 8$ as shown in Figure 4.4c could be the magic cluster observed in STM experiments.

4.3
GA Solution to the Thomson Problem

A recurring problem in computational physics and chemistry is the minimization of a structure with respect to atomic positions. One difficulty is the development of an accurate model of atomic interactions in the material. However, even once such a model is chosen, optimization is often difficult, due to the many competing structures that may be locally stable. This is especially true for noncrystalline structures such as atomic clusters and defect structures such as grain boundaries or surfaces [47]. While accurate models of materials are becoming increasingly available, and the computational time to calculate energies is rapidly decreasing, there have been relatively few developments in the optimization process. Most efforts focus on using some form of steepest-descent or conjugate gradient relaxation, or Monte Carlo or molecular dynamics simulations including simulated annealing approaches.

In this section, we show how GA can be used for investigating the long-standing Thomson problem of finding the lowest-energy configuration of n point charges on a unit sphere. The problem we consider here originated with Thomson's "plum pudding" model of the atomic nucleus. This minimization problem has been attempted by simulated annealing [48–51], Monte Carlo approaches [52,53], and symmetry considerations [54], yet none of these techniques has proven as reliable as the simplest method: a repeated random search with a steepest-descent relaxation [55,56].

The energy of n point charges constrained to lie on the surface of a unit sphere is

$$E = \frac{1}{2}\sum_{i=1}^{n}\sum_{j\neq i}\frac{1}{|\mathbf{r}_i - \mathbf{r}_j|}. \tag{4.2}$$

Even for small n, there are multiple possible stable structures; for $n \leq 20$, simulated annealing suffices to locate the global minimum [48–50]. However, this will not suffice once the number of local minima is large. The difficulty is that the number of metastable structures grows exponentially with n, and these approaches do not explore different minima sufficiently rapidly once n becomes large ($n > 70$). For $n \sim 100 - 110$, there are 50–90 metastable states; this grows to ~ 8000 for $n \sim 200$. Furthermore, for many of the structures, the basin of attraction or "catchment region" containing the global minimum is small compared with those of other minima. These difficulties are a generic feature of many systems, including the related problem of determining structures of atomic clusters [57]. Often, there are techniques to provide local optimization, such as steepest-descent or conjugate gradient algorithms. Monte Carlo simulations and simulated annealing are typically used to explore nearby minima, in an effort to improve upon the current minimum. The difficulty is that these techniques for "hopping" from one minimum to the next are time consuming, and if there are many local minima, with large barriers separating them, then these techniques are not practical. The Thomson problem is a good example of such a problem. Finding a local minimum from a random structure is straightforward, but exploring many different minima is not.

We have used a real-space GA to tackle this problem. One of the difficulties in the type of problem that we are considering is that the evaluation of the energy is time consuming, especially for problems using more accurate models of materials. For most current applications of GA, the computational effort in calculating the fitness is very small. Therefore, we cannot afford to use traditional approaches, which might require calculating the energies of thousands of structures, most of which would not be competitive [58,59].

Our approach is successful because of an interesting mating algorithm that allows for efficient exploration of different minima, while preserving the important properties of the parent structures. Unlike some previous applications of genetic algorithms [58–61], our algorithm is not based upon an artificial "genetic sequence" but is performed in real space. In the work presented here, we began with four random geometries. Using each possible pair of initial geometries, we construct 16 more candidate structures. Note that a cluster may enter in a crossover with itself, by aligning any two randomly chosen halves of the structure. From the 20 structures, we select the best four candidates, choosing only structures whose energies differ by more than $\Delta E < 10^6$ to ensure that one structure does not dominate the entire population. For $10 \leq n \leq 132$, and also for $n = 192$ and $n = 212$, we found the same minimum energies as given in Ref. [62]. Most strikingly, for $n \leq 132$, we were almost always able to find the lowest-energy structures within five generations. With these successes, we went on to search for the lowest-energy structures for $133 \leq n \leq 200$. We ran these for 10 generations, considering a total of 200 structures. Note that GA does not guarantee that the lowest energy will be found, although we believe that in most cases the final structure was the global minimum. We fitted the lowest energies to the form

$$E(n) = \frac{n^2}{2}(1 - an^{-1/2} + bn^{-3/2}). \tag{4.3}$$

The fitted values were $a = 1.10461 \pm 0.00001$ and $b = 0.137 \pm 0.001$, in reasonable agreement with the calculations of Glasser [53]. In Figure 4.8, we show the difference

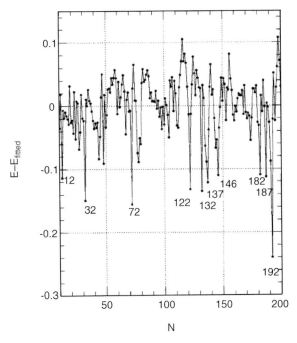

Figure 4.8 Difference between the calculated lowest-energy configuration and the fit to Equation 4.3. Note the "magic numbers" at $n = $ 12, 32, 72, 122, 132, 137, 146, 182, 187, and 192. (From Ref. [32], with permission from American Physical Society.)

between the fitted energy and the actual value for the lowest-energy structure obtained using our approach. Note that there are a series of "magic" numbers, with particularly low ground-state energies (relative to the trend given by Equation 4.3), for $n = $ 12, 32, 72, 122, 132, 137, 146, 182, and 187. In this series, the structures for $n = $ 12, 32, 72, 122, 132, and 192 have icosahedral symmetry. The icosahedral structures for $n = $ 212, 272, 282, and 312 also have very low energies [62]. Icosahedral structures have been predicted to have the lowest energy [51], but for $n = $ 42, 92, and 162, the icosahedral structures have high energies relative to the trend in Equation 4.3.

For most of the lowest-energy structures we found, the atoms tend to arrange themselves in a triangular configuration, with 12 points having five nearest neighbors and the rest having six neighbors (see Figure 4.9). With this type of configuration, the application of Euler's formula predicts that the number of faces will be $F = 2n - 4$. This prediction is confirmed for most of the lowest-energy structures, with some exceptions. The exceptions demonstrate that not all structures can be uniquely decomposed into triangles – on some structures, there are rectangular faces. This counterintuitive result illustrates the difficulties in making general statements concerning this problem. The fivefold coordinated points tend to separate themselves, suggesting that the icosahedral structures would be particularly stable, with each of the fivefold coordinated points located along a line of fivefold rotational symmetry.

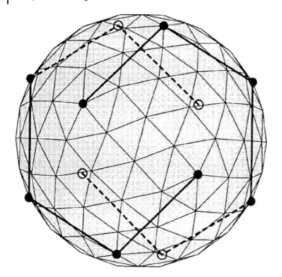

Figure 4.9 Lowest-energy structure for $n = 146$, looking down one of the twofold axes. We have emphasized the fivefold coordinated charges and indicated the interlocking C structures formed by connecting the fivefold coordinated charges to their nearest neighbor. (From Ref. [32], with permission from American Physical Society.)

The striking result is that this technique can find the lowest-energy configurations, both for the high-symmetry icosahedral structures and for structures with lower symmetry. The structures for $n = 137$, 182, and 187 are distorted icosahedral structures, with D_5 symmetry. The $n = 146$ structure, shown in Figure 4.9, has D_2 symmetry, much lower than the symmetries of the other magic numbers. Unlike many of the structures, in which the fivefold coordinated charges form equilateral triangles, the fivefold coordinated points are not in an icosahedral arrangement. Instead, the lines connecting fivefold coordinated atoms along the shortest distance between them produce two interlocking C structures. To our knowledge, no other similar structure has been predicted as being particularly favorable. We believe that there will be other magic numbers with similar structures at larger n.

It may seem surprising that such a simple GA approach works where more complicated schemes have not. We believe that there are two principal features of our technique that are important. First, we try many different geometries in parallel rather than exploring phase space in a single series of geometries. Simulated annealing or other techniques may explore several different local minima with a reasonable computational effort, but for problems with many minima these approaches become impractical. This is why a simple random search is more successful than these approaches. Second, unlike a random search, our technique of generating new structures preserves much of the previous structural optimization that has occurred. The two halves remain reasonably intact, while "healing" occurs near the joining region. Thus, while we rapidly explore other minima, we do so with a bias toward the types of low-energy structures that have already been obtained.

References

1. Deaven, D.M. and Ho, K.M. (1995) *Phys. Rev. Lett.*, **75**, 288.
2. Johnston, R.L. (2003) *Dalton Trans.*, 4193.
3. Hartke, B. (2004) *Struct. Bond.*, **110**, 33–53.
4. Wales, D.J. and Scheraga, H. (1999) *Science*, **285**, 1368.
5. Woodley, S.M. and Catlow, R. (2008) *Nat. Mater.*, **7**, 937.
6. Catlow, C.R.A., Bromley, S.T., Hamad, S., Mora-Fonz, M., Sokol, A.A., and Woodley, S.M. (2010) *Phys. Chem. Chem. Phys.*, **12**, 773.
7. Sierka, M. (2010) *Prog. Surf. Sci.*, **85**, 398.
8. Lloyd, L.D., Johnston, R.L., Roberts, C. et al. (2002) *ChemPhysChem*, **3**, 408.
9. Chuang, F.C., Wang, C.Z., and Ho, K.M. (2006) *Phys. Rev. B*, **73**, 125431.
10. Michaelian, K., Rendon, N., and Garzon, I.L. (1999) *Phys. Rev. B*, **60**, 2000.
11. Zhang, W.X., Liu, L., and Li, Y.F. (1999) *Acta Phys. Sin.*, **48**, 642.
12. Wang, J.L., Wang, G.H., and Zhao, J.J. (2003) *Phys. Rev. A*, **68**, 013201.
13. Sun, H.Q., Luo, Y.H., Zhao, J.J. et al. (1999) *Phys. Status Solidi B*, **215**, 1127.
14. Wang, B.L., Zhao, J.J., Chen, X.S. et al. (2005) *Phys. Rev. A*, **71**, 033201.
15. Hsu, P.J. and Lai, S.K. (2006) *J. Chem. Phys.*, **124**, 044711.
16. Wang, J.L., Wang, G.H., Chen, X.S. et al. (2002) *Phys. Rev. B*, **66**, 014419.
17. Bailey, M.S., Wilson, N.T., Roberts, C. et al. (2003) *Eur. Phys. J. D*, **25**, 41.
18. Diaz-Ortiz, A., Aguilera-Granja, F., Michaelian, K. et al. (2005) *Physica B*, **370**, 200.
19. Darby, S., Mortimer-Jones, T.V., Johnston, R.L. et al. (2002) *J. Chem. Phys.*, **116**, 1536.
20. Hamad, S., Catlow, C.R.A., Woodley, S.M. et al. (2005) *J. Phys. Chem. B*, **109**, 15741.
21. Wang, C., Liu, L., and Li, Y.F. (1999) *Acta Phys.-Chim. Sin.*, **15**, 143–149.
22. Roberts, C. and Johnston, R.L. (2001) *Phys. Chem. Chem. Phys.*, **3**, 5024.
23. Tomasulo, A. and Ramakrishna, M.V. (1996) *J. Chem. Phys.*, **105**, 10449.
24. Francisco, E., Martin Pendas, A., and Blanco, M.A. (2005) *J. Chem. Phys.*, **123**, 234305.
25. Ge, Y.B. and Head, J.D. (2002) *J. Phys. Chem.*, **106**, 6997.
26. Ge, Y.B. and Head, J.D. (2005) *Mol. Phys.*, **103**, 1035.
27. Hartke, B., Schutz, M., and Werner, H.J. (1998) *Chem. Phys.*, **239**, 561.
28. Guimaraes, F.F., Belchior, J.C., Johnston, R.L. et al. (2002) *J. Chem. Phys.*, **116**, 8327.
29. Deaven, D.M., Tit, N., Morris, J.R., and Ho, K.M. (1996) *Chem. Phys. Lett.*, **256**, 195.
30. Wales, D.J. and Doye, J.P.K. (1997) *J. Phys. Chem. A*, **101**, 5111.
31. Chuang, F.-C., Liu, B., Wang, C.Z., Chan, T.L., and Ho, K.M. (2005) *Surf. Sci.*, **598**, L339.
32. Morris, J.R., Deaven, D.M., and Ho, K.M. (1996) *Phys. Rev. B*, **53**, R1740.
33. Hwang, I.S., Ho, M.S., and Tsong, T.T. (1999) *Phys. Rev. Lett.*, **83**, 120; Ho, M.S., Hwang, I.S., and Tsong, T.T. (2000) *Phys. Rev. Lett.*, **84**, 5792; Hwang, I.S., Ho, M.S., and Tsong, T.T. (2002) *Surf. Sci.*, **514**, 309; Ho, M.S., Hwang, I.S., and Tsong, T.T. (2004) *Surf. Sci.*, **564**, 93.
34. Kotlyar, V.G., Zotov, A.V., Saranin, A.A., Kasyanova, T.V., Cherevik, M.A., Pisarenko, I.V., and Lifshits, V.G. (2002) *Phys. Rev. B*, **66**, 165401.
35. Jia, J.-F., Liu, X., Wang, J.-Z., Li, J.-L., Wang, X.S., Xue, Q.-K., Li, Z.-Q., Zhang, Z., and Zhang, S.B. (2002) *Phys. Rev. B*, **66**, 165412.
36. Chang, H.H., Lai, M.Y., Wei, J.H., Wei, C.M., and Wang, Y.L. (2004) *Phys. Rev. Lett.*, **92**, 066103.
37. Wang, C.Z., Pan, B.C., and Ho, K.M. (1999) *J. Phys.: Condens. Matter*, **11**, 2043.
38. Ho, K.M., Shvartsburg, A.A., Pan, B., Lu, Z.Y., Wang, C.Z., Wacker, J.G., Eye, J.L., and Jarrold, M.F. (1998) *Nature*, **392**, 582.
39. Chuang, F.C., Ciobanu, C.V., Shenoy, V.B., Wang, C.Z., and Ho, K.M. (2004) *Surf. Sci.*, **573**, L375.
40. Chuang, F.C., Ciobanu, C.V., Predescu, C., Wang, C.Z., and Ho, K.M. (2005) *Surf. Sci.*, **578**, 183; Chuang, F.C., Ciobanu, C.V., Wang, C.Z., and Ho, K.M. (2005) *J. Appl. Phys.*, **98**, 073507.
41. http://www.pwscf.org.
42. Hohenberg, P. and Kohn, W. (1964) *Phys. Rev.*, **136**, B864; Kohn, W., and Sham, L.J. (1965) *Phys. Rev.*, **140**, A1135.
43. Ceperley, D.M. and Alder, B.J. (1980) *Phys. Rev. Lett.*, **45**, 566.

44 Perdew, J.P. and Zunger, A. (1981) *Phys. Rev. B*, **23**, 5048.
45 Tersoff, J. and Hamann, D.R. (1985) *Phys. Rev. B*, **31**, 805.
46 Brommer, K.D., Needels, M., Larson, B.E., and Joannopoulos, J.D., (1992) *Phys. Rev. Lett.*, **68**, 1355; Lim, H., Cho, K., Capaz, R.B., Joannopoulos, J.D., Brommer, K.D., and Larson, B.E., (1996) *Phys. Rev. B*, **53**, 15421; Wang, Y.L., Gao, H.-J., Guo, H.M., Liu, H.W., Batyrev, I.G., McMahon, W.E., and Zhang, S.B. (2004) *Phys. Rev. B*, **70**, 073312.
47 Hoare, M.R. and McInnes, M. (1983) *Adv. Phys.*, **32**, 791.
48 Munera, H.A. (1986) *Nature*, **320**, 320.
49 Webb, S. (1986) *Nature*, **323**, 20.
50 Wille, L.T. (1986) *Nature*, **324**, 46.
51 Edmunson, J.R. (1992) *Acta Crystallogr. Sect. A*, **48**, 60.
52 Altschuler, E.L., Williams, T.J., Ratner, E.R., Dowla, F., and Wooten, F. (1994) *Phys. Rev. Lett.*, **72**, 2671;**74**, 1483 (1995).
53 Glasser, L. and Every, A.G. (1992) *J. Phys. A*, **25**, 2473.
54 Edmunson, J.R. (1993) *Acta Crystallogr. Sect. A*, **49**, 648.
55 Erber, T. and Hockney, G.M. (1995) *Phys. Rev. Lett.*, **74**, 1482.
56 Erber, T. and Hockney, G.M. (1991) *J. Phys. A*, **24**, L1369.
57 Stephen Berry, R. (1990) *J. Chem. Soc., Faraday Trans.*, **86**, 2343.
58 Venkatasubramanian, V., Chan, K., and Caruthers, J.M. (1994) *Comput. Chem. Eng.*, **18**, 833.
59 Venkatasubramanian, V., Chan, K., and Caruthers, J.M. (1995) *J. Chem. Inform. Comput. Sci.*, **35**, 188.
60 Holland, J.H. (1975) *Adaptation in Natural and Artificial Systems*, The University of Michigan Press, Ann Arbor, MI.
61 Mestres, J. and Scuseria, G.E. (1995) *J. Comput. Chem.*, **16**, 729.
62 Sloane, N.N.J.A., Hardin, R.H., Duff, T.S., and Conway, J.H. (1995) *Discrete Comput. Geom.*, **14**, 237.

5
Atomic Structure of Surfaces, Interfaces, and Nanowires

As mentioned in Chapter 1, the "natural" evolution of the field of genetic algorithm (GA) optimization should have probably been triggered after the idea of real-space representation of crossovers [1] is first clusters, then three-dimensional (3D) crystal structures, and then low-dimensional structures [i.e., two-dimensional (2D) crystals, surface reconstructions, interfaces, nanowires, etc.]. This assertion is based on the long-recognized pressing needs for solutions of structural problems in the realm of clusters and 3D crystals. In terms of relative importance, there was little actual priority given to surfaces or nanowires; however, the genetic algorithm for surfaces [2] was developed ahead of that for crystal structures [3–5]; the main reason for this "deviation" from the so-called natural evolution was, very likely, the realization that the task of finding the reconstruction of semiconductor surfaces is actually a problem of stochastic optimization [6]. As mentioned in Ref. [3], the GA for surfaces has actually contributed to its development for crystal structures by putting forth the provision that the number of atoms can be left variable if the cost function is reported in a specific way (e.g., per area as in Ref. [2] or per atom as in Ref. [3]).

In this chapter, we will show in detail the current progress in two-dimensional and one-dimensional (1D) structure optimization problems. Unlike Chapters 3 and 4, this chapter is probably nearly exhaustive as far as reviewing state of the art in low-dimensional problems solved via real-space GA optimization is concerned: this is because there are not too many 1D and 2D structural problems tackled so far by GA. The study that follows is taken, with permission from the corresponding publishers, from studies on surface reconstructions [2,7,8], interfaces [9], atomic structure of steps [10], and nanowires and nanotubes [11–13]. We believe that after the current fervor in 3D crystal structure prediction will have subsided somewhat, the GA investigations will switch focus on low-dimensional systems for several reasons:

1) These systems, especially the 2D ones, were given a tremendous boost by the discovery of graphene [14] and other nanomaterials with layered morphologies.
2) The coupling between GA and density functional theory (DFT) calculations [3–5] will have been matured to sufficient extent to obviate the use of empirical potentials. These empirical potentials are anyway mainly developed for bulk crystal structures and see very little transferability to low-dimensional structures in the nanoscale regime.

3) The idea of templating, that is, of having to use a larger system than the part subjected to GA operations in order to compute energy-related cost functions, is fairly new. This is both a novelty and a necessity, and can be exploited for problems such as the structure of dislocation cores, nanoscale inclusions, nanopores, and others.

5.1
Reconstruction of Semiconductor Surfaces as a Problem of Global Optimization

The determination of atomic structure of crystalline surfaces is a long-standing problem in surface science. Despite major progress brought by experimental techniques such as scanning tunneling microscopy (STM) and advanced theoretical methods for treating the electronic and ionic motion, the commonly used procedures for finding the atomic structure of surfaces still rely to a large extent on one's intuition in interpreting STM images. While these procedures have proven successful for determining the atomic configuration of many low-index surfaces [e.g., Si(001) and Si(111)], in the case of high-index surfaces their usefulness is limited because the number of good structural models for high-index surfaces is rather large and may not be exhausted heuristically. An illustrative example is Si(5 5 12), whose structure has been the subject of intense dispute [15–18] since the publication of the first atomic model proposed for this surface [19]. While the structure of Si(5 5 12) may still be an open problem, there are other stable surfaces of silicon such as (113) [20] and (105) that required a long time for their structures to be revealed; in the present section, we focus on Si(105) and show that there is a large number of low-energy models for this surface.

The high-index surfaces attract a great deal of scientific and technological interest since they can serve as natural and inexpensive templates for the fabrication of low-dimensional nanoscale structures. Knowledge about the template surface can lead to new ways of engineering the morphological and physical properties of these nanostructures. The main technique for investigating atomic-scale features of surfaces is STM, although, as pointed out in a recent review, STM alone is only able to provide "a range of speculative structural models which are increasingly regarded as solved surface structures" [21]. With few exceptions that concern low-index metallic surfaces [22,23], the role of theoretical methods for structural optimization of surfaces has been largely reduced to the relaxation of these speculative models. However, the publication of numerous studies that report different structures for a given high-index silicon surface (see, for example, Refs [15–19]) indicates a need to develop methodologies capable of actually searching for the atomic structure in a way that does not predominantly rely on the heuristic reasoning associated with interpreting STM data. Ciobanu and Predescu have shown that parallel tempering Monte Carlo (PTMC) simulations combined with an exponential-decay cooling schedule can successfully address the problem of finding the reconstructions of high-index silicon surfaces [6]. Such PTMC simulations, however, have a broader scope, as they are used to perform a thorough thermodynamic sampling of the surface systems under study. Given their scope, these simulations [6] are very demanding, usually requiring several tens of processors that run canonical simulations at different temperatures and exchange configurations in order to drive the low-temperature replicas into the ground state. If we focus on finding the reconstructions only at 0 K (which can be representative for crystal surfaces in the low-temperature regimes achieved in laboratory conditions), it is certainly justified to explore global optimization methods for finding the structure of high-index surfaces.

5.1.1
The Genetic Algorithm for Surface Reconstructions: the Case of Si(105)

In this section, we address the problem of finding surface reconstructions for semiconductor surfaces and describe the development of a GA for solving this problem. We also report the GA results for the Si(105) surface, which has triggered other global optimization studies of high-index surfaces. Except for the periodic vectors of the surface unit cell [which can be determined either theoretically from knowledge of the crystal structure and surface orientation or experimentally from STM or low-energy electron diffraction (LEED) measurements], no other input is necessary. An advantage of the present approach over most of the previous methodologies used for structural optimization is that the number of atoms involved in the reconstruction, as well as their most favorable bonding topology, can be found within the same genetic search. The Si(105) surface, at least in conditions of compressive strain, is known to have a single-height rebonded step structure SR [24–28]. The PTMC study [6] indicates that the SR structure is the lowest surface energy even in the absence of strain. It is interesting to note that the number of reported reconstructions for the Si(105) has increased very rapidly from 2 models [structural unit (SU) and SR] [24–26] to a total of 14 reported in Refs [6,27]. While the set of known reconstructions has expanded, the most favorable structure has remained the SR model – in contrast to the first published model [29].

Since we aim to provide an understanding of both the method and practical grounds for implementation and practice, we present the computational details wherever these details are different from what has been given in previous sections.

5.1.1.1 Computational Details for Si(105)

The simulation cell has a single-face slab geometry with periodic boundary conditions applied in the plane of the surface and no periodicity in the direction normal to it. The top atoms corresponding to a depth $d = 5$ Å (measured from the position of the highest atom) are shuffled via a set of genetic operations described below. In order to properly account for the surface stress, the atoms in a thicker zone of 15–20 Å are allowed to relax to a local minimum of the potential energy after each genetic operation. The surface slab is made of four bulk unit cells of dimensions $a\sqrt{6.5} \times a \times a\sqrt{6.5}$ ($a = 5.431$ Å is the bulk lattice constant of Si), stacked 2×2 along the [010] and [105] directions. In terms of atomic interactions, we have used the highly optimized empirical model developed by Lenosky *et al.* [30], which was

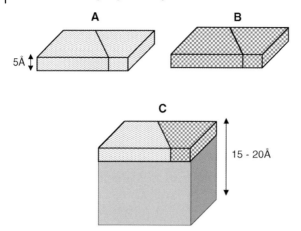

Figure 5.1 The mating operation $\mathcal{O}$: (A, B) → C. From two candidate surface structures A and B, the upper portions (5 Å thick) are separated and sectioned by the same arbitrary plane oriented perpendicular to the surface. A new slab C is created by combining the part of A that lies to the left of the cutting plane and the part of slab B lying to the right of that plane. C is placed on a thicker slab, and the resulting structure is relaxed before considering its inclusion in the pool of candidate reconstructions. (From Ref. [37], with permission from Elsevier.)

found to have superior transferability to the diverse bonding environments present on high-index silicon surfaces [6].

For the GA implementation, the "generation zero" is a pool of p different structures obtained by randomizing the positions of the topmost atoms (thickness d) and by subsequently relaxing the simulation slabs through a conjugate gradient procedure. The mating operation produces a child structure from two parent configurations as follows (refer to Figure 5.1). The topmost parts of the parent models A and B (thickness d) are separated from the underlying bulk and sectioned by an arbitrary plane perpendicular to the surface. The child structure (upper part) C is created by combining the part of A that lies to the left of the cutting plane and the part of slab B lying to the right of that plane: the assembly is placed on a thicker slab and the resulting structure C is relaxed. We have found that the algorithm is more efficient when the cutting plane is not constrained to pass through the center of the surface unit cell, and also when that plane is not too close to the cell boundaries. Therefore, we pick the cutting plane such that it passes through a random point situated within a rectangle centered inside the unit cell; numerical experimentation has shown that the algorithm performs very well if the area of that rectangle is about 80% of the area of the surface cell. We have developed two versions of the algorithm. In the first version, the number of atoms n is kept the same for every member of the pool by automatically rejecting child structures that have different numbers of atoms from their parents (mutants). In the second version of the algorithm, this restriction is not enforced, that is, mutants are allowed to be part of the pool. As we shall see, the procedure is able to select the correct number of atoms for the ground-state reconstruction without any increase over the computational effort required for one single constant-n run.

We now describe how the survival of the fittest is implemented, which is very close to the general scheme shown in Chapter 2. In each generation, a number of m mating operations are performed. The resulting m children are relaxed and considered for the possible inclusion in the pool based on their surface energy. If there exists at least one candidate in the pool that has a higher surface energy than that of the child considered, then the child structure is included in the pool. Upon inclusion of the child, the structure with the highest surface energy is discarded in order to preserve the total population p. As described, the algorithm favors the crowding of the ecology with identical metastable configurations, which slows down the evolution toward the global minimum. To avoid the duplication of members, we retain a new structure only if its surface energy differs by more than δ compared to the surface energy of any of the current members p of the pool. We also consider a criterion based on atomic displacements to account for the (theoretically possible) situation in which two structures have equal energy but different topologies: two models are considered structurally different if the relative displacement of at least one pair of corresponding atoms is greater than ε. Relevant values for the parameters of the algorithm are $10 \leq p \leq 40$, $m = 10$, $d = 5\,\text{Å}$, $\delta = 10^{-5}\,\text{meV}/\text{Å}^2$, and $\varepsilon = 0.2\,\text{Å}$. This computational procedure, down to most details, was applied for finding the structure of not only Si(105) via GA but also Si(114) and Si(337), which are described in later sections.

5.1.1.2 Results for Si(105)

The results for a Si(105) slab with 206 atoms (constant n) are summarized in Figure 5.2 a, which shows the surface energy of the most stable member of a pool of $p = 30$ candidates as a function of the number of genetic operations. The lowest surface energy starts at a very high value because the members of early generations have random surface topologies. We find that in less than 200 mating operations, the most favorable structure in the pool is already reconstructed, that is, each atom at the surface has at most one dangling (unsatisfied) bond. Furthermore, the density of dangling bonds (db) per unit area is the smallest possible for the Si(105) surface: the structure obtained is a double-step model termed DT [6] that has 4 db/$a^2\sqrt{6.5}$. The single-height rebonded structure SR is retrieved in less than 400 mating operations. The SR model is in fact the global minimum for Si(105), as found recently in an exhaustive PTMC search [6]. We have verified this finding by performing constant-n GA runs for a set of consecutive numbers of atoms, $n = 206, 205, 204$, and 203.

However, we take a further step in that we seamlessly integrate the search for the correct number of atoms with the search for the lowest-energy reconstruction: we achieve this by allowing energetically fit mutants to survive during the evolution, instead of restricting the number of atoms to be constant across the pool. The results from a GA run with variable n are shown in Figure 5.2b, in comparison with an $n = 206$ run. We notice that the variable-n evolution displays a faster drop in the lowest surface energy, as well as in the average energy across the pool. For performance testing purposes, we started the variable-n run with all the candidates having an unfavorable number of particles, $n = 204$: nevertheless, the most stable member in the pool predominantly selected a number of atoms that allows for the

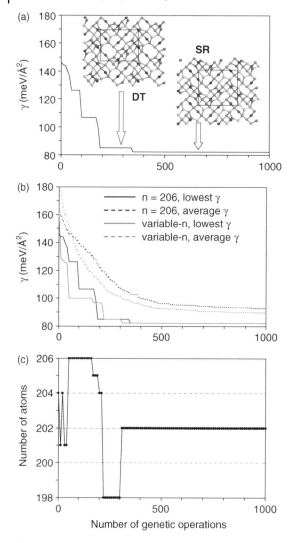

Figure 5.2 (a) Surface energy γ of the most stable Si(105) candidate from a pool of $p = 30$ structures (206-atom slabs with dimensions $a\sqrt{6.5} \times 2a$), plotted as a function of the number of mating operations. The genetic algorithm quickly retrieves the DT structure [6] and the global minimum structure SR. The insets show top views (i.e., along the $[\overline{1}05]$ direction) of the DT and SR models. The rectangles show the surface unit cells. (b) Comparison between runs with variable number ($198 \leq n \leq 210$) of atoms (red lines) and constant n, $n = 206$ (black lines). The lowest (average) surface energies are shown as solid (dashed) lines. (c) The variation of the number of atoms of the lowest-energy configuration shows that the fittest member of the pool eventually selects a value of n that is compatible with the global minimum structure SR. (From Ref. [37], with permission from Elsevier.)

5.1 Reconstruction of Semiconductor Surfaces as a Problem of Global Optimization

Table 5.1 Surface energies of 20 different Si(105) reconstructions obtained by the genetic algorithm (GA), calculated using the HOEP interatomic potential [30].

n	Surface energy (meV/Å^2)	Label from Refs [6,27]
206	82.20	SR
	85.12	DT
	88.12	DU1
	88.28	
	88.35	SU
205	86.73	DY1
	88.59	DY2
	88.61	
	88.70	
	88.97	
204	84.90	DX1
	86.04	DX2
	88.98	
	89.11	
	89.78	
203	86.52	DR1
	87.74	
	89.49	
	90.50	
	90.54	

The structures are grouped according to the number of atoms n in the simulation cell.

SR topology, that is, $n = 198, 202, 206$ (refer to Figure 5.2c). While we find no significant difference in the computational effort required by the two different evolutions, the variable-n and the constant-n ($n = 206$), the former is to be preferred: since the surface energy of a Si(105) slab is a periodic function of the number of atoms [6] with a period of $\Delta n = 4$, the variable-n run is ultimately four times faster than the sequential constant-n searches. The results from the sequential runs are summarized in Table 5.1, which shows the surface energies of 20 structures from runs with $n = 206, 205, 204$, and 203.

Motivated by recent experimental work [31] that suggests the presence of a structure with large periodic length in the [50$\bar{1}$] direction, we have also performed a GA search for configurations with larger surface unit cells ($2a\sqrt{6.5} \times 2a$), with $n = 406$ atoms. A low-energy (105) reconstruction of this size (termed DR2) was reported in our earlier work [27], where we showed that step edge rebonding lowers the surface energy below that of the DU1 model proposed in Ref. [31]. Upon annealing, the DR2 model can evolve into three structures with lower surface energies, DR2γ, DR2β, and DR2α [6]. Using the GA technique described above, we find structures that are even more favorable than DR2α (refer to Figures 5.3 and 5.4). The results displayed in Figure 5.3 indicate that the efficiency of the algorithm improves upon increasing the number of candidates in the pool, from $p = 30$ to $p = 40$. In both cases, the evolution retrieves in the same lowest-energy structure

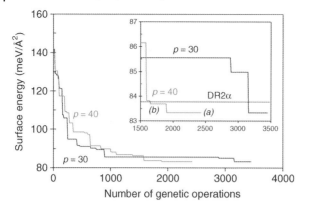

Figure 5.3 Lowest surface energy γ for pools of $p = 30$ and $p = 40$ Si(105) candidate models. The simulation slab has a periodic cell of $2a\sqrt{6.5} \times 2a$ and contains $n = 406$ atoms, of which the highest-lying 70 atoms are subject to mating operations. As seen in the inset, the procedure finds two structures (a) and (b) that are slightly more stable than the DR2α model reported in Ref. [6]; these configurations are shown in Figure 5.4, along with DR2α. (From Ref. [37], with permission from Elsevier.)

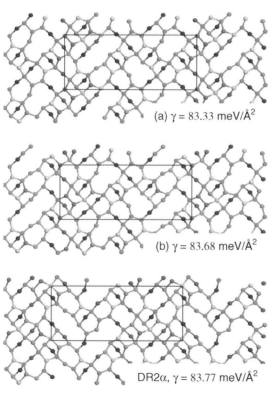

Figure 5.4 Low-energy structures of the $2a\sqrt{6.5} \times 2a$, 406-atom surface unit cells (rectangles) for Si(105). The surface energies γ are indicated next to the corresponding model labels. (From Ref. [37], with permission from Elsevier.)

with $\gamma = 83.33\,\text{meV/Å}$, which is nearly degenerate with DR2α, as the surface energy difference is only $\sim 0.44\,\text{meV/Å}^2$. It is worth noting that within the current computational resources, the variable-n algorithm performs quite well even when the number of atoms is doubled. As a test, we have run a variable-n calculation with all the structures in the pool having a number of atoms ($n = 406$), which does not correspond to the optimal configuration: this simulation still retrieves the SR model (global minimum), within about 10^4 genetic operations.

In conclusion, Ref. [2] shows that the reconstruction of semiconductor surfaces can be determined via a genetic algorithm. This procedure can be used to generate a database of model configurations for any given high-index surface, models that can be subsequently relaxed using electronic structure methods and compared with available experimental data. The process of systematically building a set of models for a given surface drastically reduces the probability of missing the actual physical reconstruction, which imminently appears when heuristic approaches are used [19,29]. The genetic algorithm presented here can naturally select the number of atoms required for the topology of the most stable reconstruction, as well as the lowest-energy bonding configuration for that number of atoms.

5.1.2
New Reconstructions for a Related Surface, Si(103)

Among the stable surfaces of Si and Ge that so far have received very little attention from a theoretical perspective are Si(103) and Ge(103). Despite the fact that they have the same orientation, experiments indicate that Si(103) and Ge(103) have very different atomic structure and morphology [32]. Ge(103) exhibits two-dimensional atomic ordering with a clear periodic pattern [33,34], while Si(103) remains rough and disordered on the atomic scale even after careful annealing [32,35]. This remarkable difference between Si(103) and Ge(103) is in itself a fundamentally interesting problem. Still, because the Si(103) surface is atomically rough and thus very difficult to tackle, so far there has not been sufficient motivation for performing extensive structure studies on this surface. This situation changed with the discovery [36] of the (103)-faceted pyramids that appear during the Si overgrowth of the Ge/Si (001) quantum dots. Motivated by the recent experiments of Wu *et al.* [36], we have set out to find atomic structure models for Si(103). Based on these models, we suggest that the rough and disordered aspect of Si(103) is due to the coexistence of several reconstructions of similar energies and different bonding topologies. Furthermore, the reconstructions presented here provide evidence that the Ge(103)-1 × 4 models previously reported [33,34] have too high surface energies to be confirmed in experiments.

The structural models for the Si(103) orientation were determined using a genetic algorithm [37] coupled with the Lenosky *et al.* [30] highly optimized empirical potential (HOEP) model of atomic interactions. We have considered three sizes of the computational cell: 1 × 2, 2 × 2, and 1 × 4, which are shown in Figure 5.5. The algorithm selects structures based on their surface energy γ, and starts with a "genetic pool" of $p = 30$ initially random atomic configurations of the surface slabs.

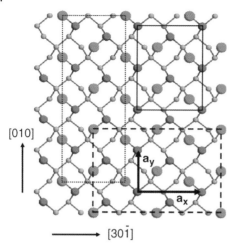

Figure 5.5 Top view of the bulk truncated Si(103) surface. The larger atoms have two dangling bonds, the medium-sized atoms have one dangling bond, and the small gray atoms are four coordinated. The unit vectors of the 1 × 1 unreconstructed primitive cell are $\mathbf{a}_x = a\sqrt{2.5}\mathbf{e}_x$ and $\mathbf{a}_y = a\mathbf{e}_y$, where $a = 5.431$ is the lattice constant of Si, and $\mathbf{e}_x$ and $\mathbf{e}_y$ are the unit vectors along $[30\bar{1}]$ and $[010]$, respectively. The rectangles show the unit cells for the 1 × 2 (solid line), the 2 × 2 (dashed line), and the 1 × 4 (dotted line) reconstructions.

The genetic pool evolves through crossover operations that combine portions of two arbitrarily chosen pool members (parents) to create another structure (child). The child structure is relaxed and retained in the pool if it is different from all existing structures and if its surface energy is sufficiently low [37]. The optimization is performed for each of the possible numbers of atoms (kept constant) that yield distinct global minima of a given surface slab. Since there are four atoms in a 1 × 2 layer (Figure 5.5), we have performed four runs for this supercell size and eight runs for each of the other two sizes, 2 × 2 and 1 × 4. The surface energies corresponding to the 600 model reconstructions retrieved are organized in the histograms shown in Figure 5.6.

To analyze the Si(103) models (Figure 5.6), we note that recent studies of high-index Si surfaces suggest that the correct (i.e., experimentally confirmed) structure either has the lowest HOEP surface energy [e.g., Si(105) in Ref. [37]] or has a surface energy that most likely lies within 3–4 meV/Å^2 from the lowest HOEP surface energy value [as in the case of Si(114) and Si(337)] [38,39]. Therefore, in order to identify good Si(103) reconstructions, we focus on a surface energy *range* that includes most of the thermodynamically favorable structures, that is, 86 meV/Å^2 < γ < 89 meV/Å^2 (refer to Figure 5.6). In this range, there are 35 models across the three periodicities considered (Figure 5.5). Of these models, 32 are distinct in the sense that the large period structures (1 × 4 and 2 × 2) cannot be reduced to the repetition of a single 1 × 2 model.

From the 32 distinct structures, we have identified a few pairs of configurations that exhibit minor differences such as bonds relaxing to slightly different local

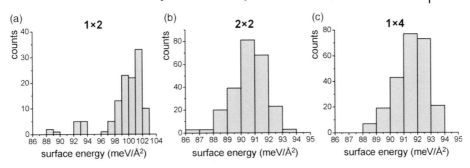

Figure 5.6 Histograms of the surface energies retrieved by the genetic algorithm for the (a) Si(103)-1 × 2, (b) Si(103)-2 × 2, and (c) Si(103)-1 × 4 reconstructions.

minima but otherwise making up the same topology at the surface. More notably, there are also groups of nearly degenerate reconstructions with markedly different atomic bonding, but with nearly equal (and low) surface energies. Some of these reconstructions are depicted in panels (a)–(e) of Figure 5.7. We have found that the atomic-scale features that appear frequently on most of the favorable reconstructions (not only those shown in Figure 5.7) are the dimers and the rebonded atoms, which would be expected for stepped Si(001) surfaces. Dimers and rebonded atoms occur in a wide variety of relative configurations for any of the low-energy Si(103) reconstructions. Interestingly, the dimer-rebonded atom configuration that is solely responsible for the lowest-energy structure of Si(105) [40,41] is also encountered on Si(103); this configuration is made up of two rebonded atoms that "bridge" at the base of a dimer to form a shape that resembles somewhat the letter u [42]. Figure 5.7 shows one such u motif marked in black in panel (b), which can readily be spotted in the other panels as well. Another known motif that appears (though not as frequently as the u) on the low-energy Si(103) reconstructions is the tetramer [43], denoted by t in Figure 5.7c.

The similarity between the best Si(103) reconstruction found here (Figure 5.7b) and the single-height rebonded (SR) model [40] for Si(105) is quite striking, as both models have two u motifs in their respective unit cells and nearly equal density of dangling bonds, that is, $1.58 \, db/a^2$ for Si(103) versus $1.57 \, db/a^2$ for SR. Since the unit cells of Si(105)-1 × 2 and Si(103)-2 × 2 have different sizes, the best Si(103) model allows for an efficient arrangement of its u motifs at the cost of introducing additional surface stress. Therefore, the resulting lowest surface energy for Si(103) (Figure 5.7b), $86.45 \, meV/Å^2$, is higher than the surface energy of the SR model [37], $82.20 \, meV/Å^2$.

The surface stress associated with low-energy Si(103) structures is tensile, because most of the bonds are stretched in order to achieve a low dangling bond density. On the other hand, a very large number of dangling bonds per area increase the surface energy even though the atoms at the surface would have significantly more room to relax. This is the case of the reconstruction proposed originally for the Ge(103)-1 × 4 surface [33,34], which is shown in Figure 5.7f after scaling to the lattice constant of Si and relaxation at the HOEP level. The surface energy of the model in Figure 5.7f is

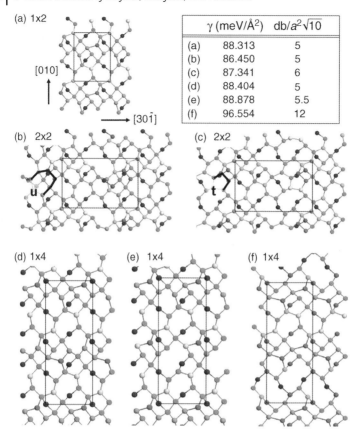

Figure 5.7 Model reconstructions (top views) for the Si(103) surface (a)–(f). Structures (a)–(e) have been obtained in this work, while structure (f) shows the model previously proposed for the (103) orientation [33,34]. The periodic cell is marked by a rectangle in each case. The inset shows the surface energy γ of models (a)–(f) calculated using the Lenosky et al. [30] potential and their number of dangling bonds (db) per 1×2 unit area.

96.55 meV/Å^2, clearly larger than the surface energy of *any* of the 240 structures accounted for in Figure 5.6c. Even though the main focus of this chapter is not on Ge(103), we were intrigued by finding such a high surface energy for model (f), so we recalculated the surface energies of all 1×4 models using an empirical potential for Ge [44]. We have found a surface energy of 91.67 meV/Å^2 for model (f) with the Tersoff potential [44], while the surface energies of all other 1×4 structures (scaled to the lattice constant of Ge) ranged between 85.94 and 95.02 meV/Å^2. This finding suggests that a reevaluation of the accepted Ge(103)-1×4 model [33,34] may be warranted in the near future.

We conclude with a short discussion of the physical implications of having a large number of low-energy reconstructions available for the Si(103) surface.

The existence of multiple models with similar surface energies but with very different topologies and different spatial periodicities suggests that it is possible for such models to *coexist* on the Si(103) orientation, a proposal that has recently been made for the case of Si(105) as well [45]. Indeed, experiments to date [35] show that both Si(103) and Si(105) are atomically rough and exhibit no discernible two-dimensional periodicity even after careful annealing. The proposal that several structural patterns can coexist on the same nominal orientation would have little value if any two models placed next to one another on the (103) surface were to give rise to domain boundaries with very high formation energies. However, we have found that *different* 1×2 models do indeed appear next to one another without substantially increasing the surface energy of the reconstructions with larger unit cells; refer, for example, to Figure 5.7e, in which the 1×2 model (a) occupies the upper half of the 1×4 cell, while the lower half has a different structure. Since there exists a vast array of energetically favorable motifs made of dimers and rebonded atoms, entropy considerations also support the idea of various structural patterns coexisting on the Si(103) surface.

In summary, we have used a genetic algorithm to find Si(103) reconstructions and proposed that the atomic-scale roughness experimentally observed for this surface is due to the coexistence of several nearly degenerate structural models with different bonding topologies and surface periodicities but with similar surface energies. We have found that the low-energy (103) reconstructions largely display the same atomic-scale motifs (combinations of dimers and rebonded atoms) as Si(105) [45], which has led us to believe that the physical origin of the observed disorder [35] is the same for both Si(103) and Si(105). In the case of Si(105), the structural degeneracy is lifted upon applying compressive strain [45] or through the heteroepitaxial deposition of Ge [40]. For Si(103), it was shown that low coverages of indium can result in the emergence of a preferred reconstruction pattern [46]. The possibility to remove the degeneracy and create a periodic pattern on Si(103) by epitaxially depositing Ge at *low coverage* has not been investigated [47]. If such experiments were to be performed, the calculations presented here would predict that the most likely model to emerge is that in Figure 5.7b, which is similar to the SR reconstruction of Ge/Si(105) [40]. Upon comparing the structures retrieved by the genetic algorithm with the existing model [33,34] for the Ge(103) surface, we have found that the latter has a density of dangling bonds that is 2.4 times larger than that of the best (103) models. The models presented here, we hope, can play an important role in revisiting the currently accepted structure of Ge(103), as well as in explaining the (103)-faceted islands [36] that appear upon Si capping of the Ge/Si(001) quantum dots.

5.1.3
Model Reconstructions for Si(337), an Unstable Surface: GA Followed by DFT Relaxations

Although unstable, the Si(337) orientation has been known to appear in diverse experimental situations such as the nanoscale faceting of Si(112) or in the case of miscutting a Si(113) surface. Various models for Si(337) have been proposed over time, which motivate a comprehensive study of the structure of this orientation. Such a study is undertaken in this chapter, where we report the results of a genetic algorithm optimization of the Si(337)-(2 × 1) surface. The algorithm is coupled with a highly optimized empirical potential for silicon, which is used as an efficient way to build a set of possible Si(337) models; these structures are subsequently relaxed at the level of *ab initio* density functional methods. As we shall see, the optimization retrieves most of the previously proposed models [48–51], as well as a number of other reconstructions that could be relevant for experimental situations where Si(337) nanofacet arises – other than the two (337) unit cells that appear as part of the (5 5 12) structure.

The *unstable* high-index orientations are important in their own right, as they often give rise to remarkable periodic grooved morphologies that may be used for the growth of surface nanowires. The wonderful morphological and structural diversity of high-index surfaces can be appreciated, for instance, from the work of Baski et al. [52], who investigated systematically the surface orientations between Si (001) and Si(111). An interesting case of unstable surface is Si(337), whose atomic structure is the subject of this chapter. As a historical account, we note that the Si(337) orientation was concluded to be stable in early studies by Ranke and Xing [53], Gardeniers et al. [54], and Hu et al. [48]. This conclusion about the stability of Si(337) was consistent with the reports of Baski and Whitman, who showed that Si(112) decomposes into quasi-periodic Si(337) and Si(111) nanofacets [55] and that Si(337) is a low-energy orientation [56]. However, further studies by Baski and Whitman revealed that Si(337) itself facets into Si(5 5 12) and Si(111) [57], demonstrating that Si(337) was, in fact, unstable. Since Si(337) is not stable and Si(5 5 12) is [49,58], the clean Si(112) orientation should facet into Si(5 5 12) and Si(111). However, as explained in Ref. [57], nanofacets are too narrow to form complete Si(5 5 12) units and mostly settle for the nearby Si(337) orientation. A more recent report [59] shows another situation where Si(337) rather than the expected Si(5 5 12) arises. For a Si(113) surface that is miscut toward [111], the large (5 5 12) reconstruction has not always been found. Instead, a Si(337) phase has been seen to coexist with Si(113) terraces [59]. The high-resolution transmission electron microscope (HRTEM) images in Ref. [59] leave no doubt that the periodicity of these nanofacets corresponds to the (337) orientation. The reason for the quasi-stability of Si(337) reported in earlier works [53,54,48] could be the curvature or the size of substrates used in those investigations; this explanation is also consistent with Ref. [59], where the issue of sample curvature is acknowledged.

With one exception [48], atomic-scale models for Si(337) were not sought separately, but rather as structural parts of the Si(5 5 12)-(2 × 1) reconstruction [49–51,60]. As shown in Ref. [49], the Si(5 5 12) unit cell consists of two Si(337) units and one Si(225) unit. The first model reconstruction proposed for Si(5 5 12) [49] appeared somewhat corrugated when viewed along the [$\bar{1}$10] direction; so did a second model by Ranke and Xing [60]. On the other hand, HRTEM measurements [50] showed a flatter profile for Si(5 5 12), and different model reconstructions were proposed [50,51] in order to account for the observed flatness. Total energy density functional calculations for the energy of Si(5 5 12) have only recently been published, which suggest that the latest models [50,51] have lower energies than the earlier proposals [49,60].

5.1.3.1 Results for Si(337) Models

The computational details are the same as in the Si(114) surface, with the obvious exception of the dimensions and periodic vectors of the supercell as they correspond to Si(337). There are four possibilities in terms of the number of atoms in the slab that yield distinct global energy minima of the Si(337) periodic cell. This has been determined by performing constant-n genetic algorithm optimization for computational slabs with consecutive numbers of atoms $n (264 \leq n \leq 272)$ and identifying a periodic behavior of the lowest surface energy as a function of n. This procedure of determining the symmetry distinct numbers of atoms in the slab has been detailed in Ref. [6]. The HOEP global minima as well as selected local minima of the surface energy for different numbers of atoms are summarized in Table 5.2, along with the density of dangling bonds per unit area and the surface energy computed at the DFT level.

As a general comment, the DFT energy ordering does not coincide with that given by HOEP, indicating that the transferability of HOEP to Si(337) is not as good as in the case of Si(001) and Si(105); to cope with this transferability issue, we have considered

Table 5.2 Surface energies of selected Si(337)-(2 × 1) reconstructions, sorted by the number of atoms n in the periodic cell.

n	Bond counting (db/$a^2 \sqrt{16.75}$)	HOEP (meV/Å^2)	DFT (meV/Å^2)	Figure
266	10 (10)	87.37	109.04	
	10 (10)	87.74	97.22	5.9b
	6 (4)	87.79	100.39	5.9c
	10 (10)	89.60	102.35	
	10 (10)	89.79	101.58	5.9e
	(10)	—	101.16	5.9f
	4 (3)	92.07	94.47	5.9a
	4 (4)	93.87	101.43	5.9d
267	8 (8)	92.35	99.95	
	8 (8)	92.37	96.47	
	10 (8)	92.47	107.72	
	9 (8)	99.11	101.33	
268	8 (6)	81.99	93.13	5.10c
	8 (8)	83.11	95.17	
	8 (6)	83.43	89.61	5.10b
	8 (8)	84.90	94.52	5.10d
	8 (6)	85.47	90.19	
	8 (6)	85.94	89.12	5.10a
269	4 (4)	89.18	93.50	5.11a
	6 (6)	92.58	99.21	5.11b

The second column shows the number of dangling bonds per unit area, counted after relaxation with HOEP; the dangling bond density at the DFT level is shown in parentheses. Columns three and four list the surface energies given by the HOEP potential [30] and by density functional calculations [61].

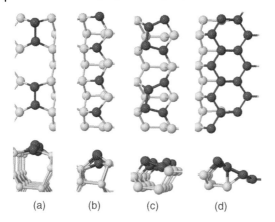

Figure 5.8 Structural features (top and side views) that can be present on low-energy Si(337) reconstructions. (a) Dimers. (b) Rebonded atoms. (c) tetramers. (d) Honeycombs. In each case, the atoms that make up the motif are shown in darker shade. (From Ref. [7], with permission from American Institute of Physics.)

more local minima (than listed in Table 5.2) when performing DFT relaxations, as mentioned in Chapter 2. In the terminology of potential energy surface (PES) theory, we first sample the main basins using the genetic algorithm coupled with HOEP and then recalculate the energetic ordering of these basins using DFT. From Table 5.2 it is apparent that the least favorable number of atoms is $n = 267$ (modulo 4) at both the HOEP and DFT levels. Therefore, we focus on describing here the reconstructions that have numbers of atoms $n = 266, 268$, and 269.

For $n = 266$, the best model that we obtained at the DFT level is made of a fused assembly of dimers, tetramers, and slightly puckered honeycombs, followed (in the $[77\bar{6}]$ direction) by a row of rebonded atoms. These individual atomic-scale motifs have been reported previously for different surfaces [49,51,62], and depicted for convenience in Figure 5.8. Their complex assembly shown in Figure 5.9a has a surface energy of 94.47 meV/Å^2; the reconstruction in Figure 5.9a has a corrugated aspect when viewed from the $[\bar{1}10]$ direction, which may account for its relatively high surface energy (refer to Table 5.2). Higher-energy models with $n = 266$ are illustrated in Figure 5.9b–d and have surface energies ranging from 2.75 to about 7 meV/Å^2 above the surface energy of the model in Figure 5.9a. Dimers (D) and tetramers (T) can be clearly identified in model in Figure 5.9b, while configuration in Figure 5.9c is characterized by the presence of honeycombs (H) and rebonded atoms (R). In Figure 5.9d, the only features from the set in Figure 5.8 that can be separately identified are the rebonded atoms R. Nearly degenerate with structure in Figure 5.9d, we find two other models (Figure 5.9e and f, with different tilting of the dimers belonging to tetramer groups) that contain six-member ring π-chains [49]. The six-member rings are labeled by 6-r in Figure 5.9e and f and are supported by tetramer-like features (denoted by t in Figure 5.9) whose dimers are made of fully coordinated atoms.

The optimum number of atoms is $n = 268$, as indicated in Table 5.2 at both HOEP and DFT levels – coincidentally, this is also the number of atoms that corresponds to

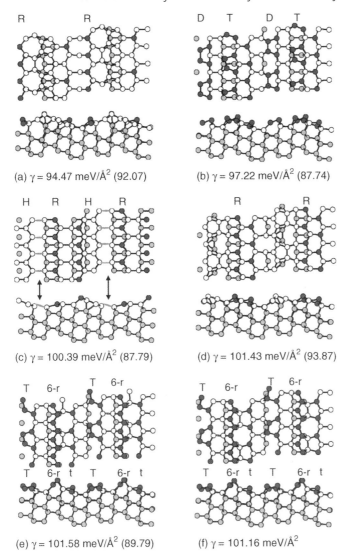

(a) $\gamma = 94.47$ meV/Å^2 (92.07)

(b) $\gamma = 97.22$ meV/Å^2 (87.74)

(c) $\gamma = 100.39$ meV/Å^2 (87.79)

(d) $\gamma = 101.43$ meV/Å^2 (93.87)

(e) $\gamma = 101.58$ meV/Å^2 (89.79)

(f) $\gamma = 101.16$ meV/Å^2

Figure 5.9 Models of Si(337) reconstructions with $n = 266$ atoms per (2×1) unit cell, after DFT relaxation (top and side views). The surface energy computed from first-principles is indicated for each structure, along with the corresponding value (in parentheses) determined with the HOEP interaction model [30]. The dark shade marks the undercoordinated atoms. The four-coordinated atoms that are exposed at the surface are shown in white. Apart from the relaxations, dimer tilting, and perhaps the relative phase of dimerization, the structure shown in panel (b) is the same as that proposed in Ref. [48] and reconstructions in panels (e) and (f) are similar to the model in Ref. [49]. (From Ref. [7], with permission from American Institute of Physics.)

two complete (337) bulk truncated primitive cells. We have further verified this number by performing a variable-n genetic algorithm started with all the members in the pool initially having $n = 267$ atoms. The algorithm has found the correct number of atoms ($n = 268$) and the HOEP global minimum in less than 1000 operations. While different starting configurations may change this rather fast evolution toward the global minimum, our tests indicate that for simulation cells of similar size several thousand genetic operations are sufficient when using the implementation described in Ref. [2]. The lowest-energy model as given by DFT is shown in Figure 5.10a and consists of dimers D, rebonded atoms R, and

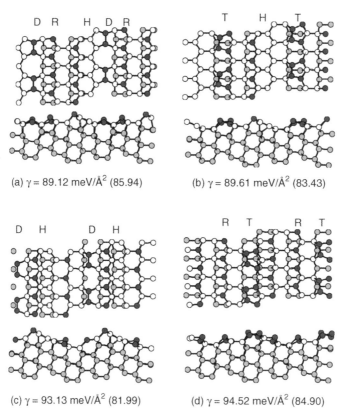

(a) $\gamma = 89.12$ meV/Å^2 (85.94)

(b) $\gamma = 89.61$ meV/Å^2 (83.43)

(c) $\gamma = 93.13$ meV/Å^2 (81.99)

(d) $\gamma = 94.52$ meV/Å^2 (84.90)

Figure 5.10 Si(337) reconstructions with $n = 268$ atoms per (2×1) unit cell, after DFT relaxation (top and side views). The surface energy computed from first-principles is indicated for each structure, along with the corresponding value (in parentheses) determined with the HOEP interaction model [30]. The dark shade marks the undercoordinated atoms, while the four-coordinated atoms that are exposed at the surface are shown in white. The lowest-energy Si(337)-(2×1) reconstructions are shown in panels (a) and (b). These structures differ in the position of the dimers in the unit cell: the dimers D can be part of a seven-member ring [side view, panel (a)], or part of a tetramer T [panel (b)]. The models shown in panels (a) and (b) are the same as those reported in Ref. [51] for the two (337) units that are part of a (5 5 12) cell. (From Ref. [7], with permission from American Institute of Physics.)

honeycombs H, in this order along [776]. We note that the number of dangling bonds is smaller at the DFT level (refer to Table 5.2): the reason for the dangling bond reduction is the flattening of the honeycombs (not planar at the level of empirical potentials), which creates two additional bonds with the subsurface atoms. The genetic algorithm search retrieves another low-energy model, in which the dimers are displaced toward and combine with the rebonded atoms, forming tetramers T (Figure 5.10b). The two models in Figure 5.10a and b are nearly degenerate, with a surface energy difference of $\sim 0.5\,\text{meV}/\text{Å}^2$ relative to one another. A different number and ordering of the structural motifs that are present in Figure 5.10a give rise to a noticeably larger surface energy. Specifically, the arrangement of dimers D and honeycombs H (Figure 5.10c) increases the surface energy by $4\,\text{meV}/\text{Å}^2$ relative to the model in Figure 5.10a. An even higher-energy reconstruction made up of tetramers and rebonded atoms is shown in Figure 5.10d.

When the number of atoms in the simulation cell is $n = 269$, we find that the ground state is characterized by the presence of pentamers, subsurface interstitials, and rebonded atoms (refer to Figure 5.11a). Interestingly, this structure is nearly flat, with the rebonded atoms being only slightly out of the plane of the pentamers. The pentamers are "supported" by six-coordinated subsurface interstitials. For the Si(337) cell with $n = 269$ atoms, we find a very strong stabilizing effect of the pentamer–interstitial configuration: the next higher-energy structure (Figure 5.11b) is characterized by adatoms and rebonded atoms, with a surface energy that is $\sim 5\,\text{meV}/\text{Å}^2$ higher than that of the ground state in Figure 5.11a. We note that subsurface interstitials have long been reported in the literature to have a stabilizing effect on the reconstruction of Si(113) [63]. Next, we compare the models described above with previously published works on Si(337) structure.

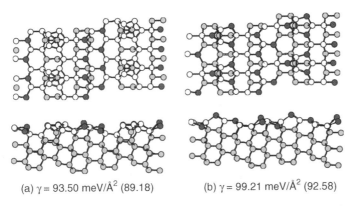

(a) γ = 93.50 meV/Å² (89.18) (b) γ = 99.21 meV/Å² (92.58)

Figure 5.11 Models of Si(337) reconstructions with $n = 269$ atoms per (2×1) unit cell, after DFT relaxation (top and side views). The surface energy computed from first-principles is indicated for each structure, along with the corresponding value (in parentheses) determined with the HOEP interaction model [30]. The dark shade marks the undercoordinated atoms, while the four-coordinated atoms that are exposed at the surface are shown in white. The best $n = 269$ model [panel(a)] is stabilized by flat pentamers and subsurface interstitials. (From Ref. [7], with permission from American Institute of Physics.)

5.1.3.2 Discussion

To our knowledge, the first model of the Si(337) surface was proposed by Baski *et al.* [49] in their original work on the structure of Si(5 5 12). The model contained surface tetramers and six-member ring π-chains and was similar to the structures depicted in Figure 5.9e and f. We observed a strong relaxation of the π-bonded chain, in which one of the atoms of each six-member ring protrudes out of the surface and settles well above its neighbors. The surface energy of the model in Figure 5.9f is not particularly favorable, although it is the lowest of all the models containing six-member ring π-chains that we have investigated. Heuristically speaking, if the undercoordinated atoms of the six-member rings are removed (four atoms per (2 × 1) unit cell), then one would obtain the model of Hu *et al.* [48], which is made of dimers and tetramers. The genetic algorithm with $n = 266$ (modulo 4) has also retrieved this model, which is shown in Figure 5.9b after DFT relaxation.

Models in Figure 5.9b,e, and f have a high density of dangling bonds per unit area, $10\,\text{db}/a^2\sqrt{16.75}$. While the dangling bond density may seem to account for the large surface energy of these models, a close inspection of Table 5.2 shows that there is, in fact, no clear correlation between the dangling bonds and the surface energy. For instance, at the same number of unsaturated bonds per area, models in Figure 5.9b, e, and f span a $4\,\text{meV}/\text{Å}^2$ range in surface energy. Furthermore, if we take the DFT model in Figure 5.9a as reference, we note that an increase by 1 db per unit cell (Figure 5.9a–c) results in a surface energy increase of $\sim 7\,\text{meV}/\text{Å}^2$; on the other hand, an increase of 7 db in the simulation cell (Figure 5.9a and b) increases the surface energy by less than $3\,\text{meV}/\text{Å}^2$. Similar arguments for the lack of correlation between missing bonds and surface energy can be made by analyzing structures with other numbers of atoms in Table 5.2. At the optimum number of atoms ($n = 268$), we find that with the same dangling bond density, models in Figure 5.10a–c span a $4\,\text{meV}/\text{Å}^2$ surface energy range: this range is so large that it includes the global minimum as well as higher surface energy local minima that may not be observable in usual experiments. These findings are consistent with the statement that the minimization of surface energy is controlled not only by the reduction of the dangling bond density but also by the amount of surface stress caused in the process [49]. Given that the heuristic approaches may account only for the dangling bond density when proposing candidate reconstructions, the need for robust minimization procedures [2,6] becomes apparent.

A notable configuration with $4\,\text{db}/a^2\sqrt{16.75}$ is shown in Figure 5.9c, where the bonds that join the (lower corner of) honeycombs and the neighbors of the rebonded atoms are markedly stretched beyond their bulk value. The stretched bonds (indicated by arrows in Figure 5.9c) cause the surface energy to become very high, $100.39\,\text{meV}/\text{Å}^2$ at the DFT level. Interestingly, the surface stress (and surface energy) is very efficiently lowered by adding dimers such that stretched bonds in Figure 5.9c are broken, and the dimers "bridge" over as shown in Figure 5.10a. The addition of one dimer per simulation cell results in the most stable Si(337) reconstruction that we found, with $\gamma \approx 89\,\text{meV}/\text{Å}^2$. A structure similar to that in Figure 5.10a was previously proposed by Liu *et al.* [50] to explain their HRTEM data. These authors also proposed a possible configuration for the other unit of (337) that is part of the large Si(5 5 12) unit cell, which has essentially the same bonding topology as in Figure 5.10b. Although no atomic-scale calculations were performed in Ref. [50], the two (337) models of Liu et al. were in qualitative agreement with more recent DFT results by Jeong et al. [51], as well as with STM and HRTEM images. When performing *ab initio* relaxations, we found that the HOEP models that corresponded most closely to those in Ref. [50] end up relaxing into the configurations put forth by Jeong et al. [51]. The structures shown in Figure 5.10a and b allow (337) nanofacets to intersect (113) facets without any bond breaking or rebonding. This absence of facet edge rebonding, as well as the relatively low energy of Si(337), gives rise to short-range attractive interactions between steps on miscut Si(113) [64], which is consistent with the experimental observation of (337) step-bunched phases [59].

5.1.4
Atomic Structure of Steps on High-Index Surfaces

One-dimensional nanostructures presently show tremendous technological promise due to their novel and potentially useful properties. For example, gold chains on stepped silicon surfaces [65,66] can have tunable conduction properties and rare-earth nanowires and bismuth nanolines have unusual straightness and length [67] and can thus be useful as nanoscale contacts on chips or as templates for the design of nanodevices. The structure of steps on silicon surfaces is of key interest, as it can help trigger a step-flow growth mode [68] useful for preparing high-quality wafers. Understanding the formation, properties, and potential applications of these intriguing 1D nanostructures requires knowledge of the atomic positions of various possible adsorbate species, as well as of the location of the silicon atoms at the step edges.

Motivated by the need to find good candidates for one-dimensional structures on surfaces, we have set out to develop a global search procedure that creates and selects atomic models based on their formation energy. To fix ideas, we tackle here the following problem: Given a stable surface orientation with a known reconstruction and given a direction in the plane of this surface, find the atomic structure of the steps oriented along that direction. We report a robust and generally applicable variable-number genetic algorithm for determining the atomic configuration of crystallographic steps and exemplify it by finding structures for several types of monoatomic steps on Si(114)-2 × 1. We show that the location of the step edge with respect to the terrace reconstructions, the step width (number of atoms), and the positions of the atoms in the step region can all be simultaneously determined.

5.1.4.1 Supercell Geometry and Algorithm Details

Focusing on the case of steps on the Si(114)-2 × 1 surface, experiments show that straight steps form along the $[\bar{1}10]$ and the $[22\bar{1}]$ directions [69], which are precisely the directions of spatial periodicity of the reconstructed Si(114) unit cell [62]. For each of the two directions, we can define two types of steps (up and down), and for each step type, there are two relative positions of the reconstructed unit cells on the upper and lower terraces: one in which the unit cells on terraces are in registry (normal) and another in which they are offset along the step direction (shifted). A crystallographic analysis of the Si(114) surface shows that of the four possible terrace

(a) Top view, along [1̄ 1̄ 4]

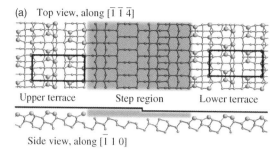

Upper terrace   Step region   Lower terrace

Side view, along [1̄ 1 0]

(b) Parent A   Child C   Parent B

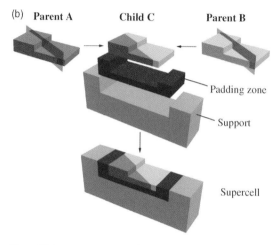

Padding zone

Support

Supercell

Figure 5.12 (a) Step region (shaded) for which the number of atoms, their positions, and the location of the step edge are to be determined. The region shown corresponds to the [1̄10]-down step configuration and is surrounded by reconstructed terraces with the unit cell marked by rectangles. (b) Crossover operation through which the genetic pool of step structures is evolved. The step energies used in the selection process are computed after relaxing a padding zone (in addition to the step region) with the support kept fixed. (From Ref. [10], with permission from American Physical Society.)

configurations for each step direction, there can be only two that are topologically distinct. The configurations that we have to address are therefore four, denoted here by [1̄10]-down, [1̄10]-up, [22$\bar{1}$]-normal, and [22$\bar{1}$]-shifted. Figure 5.12a illustrates the [1̄10]-down configuration, while the remaining ones are described in the following.

The reconstructed unit cell on terraces has dimensions of $3a \times a\sqrt{2}$, where $a = 5.431$ Å is the bulk lattice constant of Si. The height of monoatomic steps on Si (114) is $h = \sqrt{2}a/12$. The down and up [1̄10] steps create intrinsic step widths $\Delta = -11a/6$ and $\Delta = -7a/6$, respectively. In order to correctly apply (nonorthogonal) periodic boundary conditions [70], the terrace must be lowered or raised by an amount equal to h and displaced along the y-direction (i.e., along the step) by $s_{\parallel} = \sqrt{2}a/4$ upon any periodic translation of length L_x performed in the direction perpendicular to the step that lies in the plane of the flat Si(114) terrace. The supercell dimensions corresponding to the $[\bar{1}10]$ steps are $L_x = 3ak + \Delta$ and $L_y = \sqrt{2}a$, where k is an integer set to be large enough ($k = 12$) that the elastic repulsion between the periodic images of the step is negligible. Steps in the $[22\bar{1}]$ direction can differ only through the relative positioning of the reconstruction pattern on the upper and lower terraces. We found two such relative positions denoted as *normal* (for which $\Delta = -\sqrt{2}a/4$) and *shifted* ($\Delta = -3\sqrt{2}a/4$). The translation along the step direction for both of them is $s_{\|} = 7a/6$, and the dimensions of the supercell are $L_x = k\sqrt{2}a + \Delta$ and $L_y = 3a$.

Referring to Figure 5.12b, the crossover of parents A and B is achieved by sectioning them with the same random plane and then retaining atoms from each parent located on different sides of this plane to create the child C. The plane is chosen here to be parallel to [114] with any azimuthal angle about this direction allowed and passes through a random point in the rectangle projected by the step zone onto the (114) plane. The operation so defined (Figure 5.12b) has built-in potential to generate child structures with different numbers of atoms from their parents. Any child that is *structurally distinct from all pool members* is considered for inclusion in the genetic pool based on its formation energy per unit length, which should be lower than that of the highest-ranked (i.e., least favorable) member of the pool. To preserve the total population, the structure with the highest formation energy is discarded upon inclusion of a child. In a genetic algorithm run, the crossover operation is repeated to ensure that the lowest-energy structure of the pool has stabilized; as such, the present systems require on the order of 2000 operations.

The formation energy of a step structure is defined as a per-length quantity that is in excess to the bulk and surface energies [70], and therefore can be written as

$$\Lambda = \frac{1}{L_y}(E_m - N_m e_b - \gamma A), \tag{5.1}$$

where E_m is the total energy of the N_m atoms that are allowed to move within a projected area $A = L_x L_y$ with the dimension L_x (L_y) perpendicular (parallel) to the step, e_b is the bulk cohesion energy of Si, and γ is the surface energy of the flat Si (114) surface. The potential we have chosen to model the atomic interactions is the one developed by Lenosky *et al.* [30] because it has shown reasonable transferability for diverse atomic environments present on high-index Si surfaces. If all the atoms of the supercell are allowed to move when calculating the formation energy (Equation 5.3), then each update of the genetic pool would be too slow for the algorithm to be practical. On the other hand, if we relax the atoms only in the step zone, then Equation 5.3 would include not only the formation energy but also the elastic interactions of the step with the nearby rigid boundaries of the step region. To reach a good compromise between the full accuracy of Equation 5.3 (which would be achieved when *all* atoms in the supercell are relaxed) and the speed required to sort out many structures per unit time, we introduce a padding zone that is relaxed along with the step region while keeping the reconstructed support zone fixed (refer to Figure 5.12b). At the end of any genetic algorithm run, a full relaxation (all atoms allowed to move) is performed for all pool members in order to refine their step energies and structures.

5.1.4.2 Results for Step Structures on Si(114)

Typical results of the genetic algorithm for steps are shown in Figure 5.13a, which displays the evolution of the lowest and of the average formation energy of the genetic pool as a function of the number of crossover operations. Both the lowest

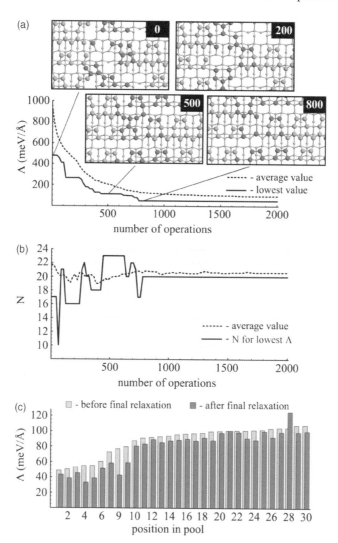

Figure 5.13 Finding the structure of [22$\bar{1}$]-normal steps on Si(114). (a) Step energy Λ of the lowest-energy structure (solid line) and averaged across the pool (dashed line) during the genetic evolution. The lowest-energy structure is shown after 0, 200, 500, and 800 crossover operations; the atoms subjected to optimization are shown as darker spheres in the insets. (b) Evolution of the average number of atoms across the pool (dashed line) and of the atom number corresponding to lowest-energy member (solid line). (c) Step energies before and after the final relaxation of all members of the genetic pool. The formation energies decrease upon full relaxation unless bonds are broken in the process (as found in the case of structure No. 28). (From Ref. [10], with permission from American Physical Society.)

and the average formation energies show a rapid decrease at the beginning of the evolution, followed by a much slower decay in the later stages. The lowest-energy [22$\bar{1}$]-normal configuration was found in less than 1000 operations, and has been retrieved in four runs started from different generation zero scenarios with no significant change in the total number of crossover moves. Since the crossover operation described above creates structures with variable numbers of atoms, the number of atoms N in the step region is optimized at the same time as the atomic positions (N is always smaller than the total number of atoms allowed to relax, N_m). Figure 5.13b shows the search for the optimal particle number N, displaying the evolution of the average number of atoms in the genetic pool and the particle number corresponding to the lowest-energy member. Figure 5.13c shows that upon final full relaxation, a certain amount of energetic reordering occurs, but this reordering does not eliminate from consideration any of the structures deemed favorable prior to the full relaxation. When using the algorithm for an arbitrary line defect, the formation energy comparison (before and after final relaxation) offers the most useful criterion for adjusting the size of the padding zone so as to provide sufficient relaxation without rendering the calculations intractable.

The best three structures found for the up- and down-steps oriented along [$\bar{1}$10] are shown in Figure 5.14, along with their formation energies after the final relaxation and their optimal atom numbers N. The most favorable up- and down-steps both have negative formation energies, which is a known artifact of the empirical potentials

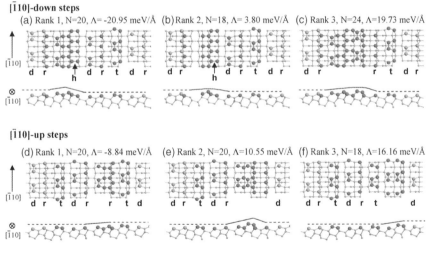

Figure 5.14 Low-energy step structures of [$\bar{1}$10]-oriented steps on the Si(114) surface (a through f). The atoms subjected to optimization are represented by dark spheres, while the atoms making up the terrace reconstructions are the lighter ones. The remaining atoms are shown as smaller gray spheres. The structural motifs on the terraces are rows of dimers (d), rebonded atoms (r), and tetramers (t). Some or all of these motifs also make up the shown step structures with the exception of the most favorable down-step models (a and b) that include hexagon rows denoted by "h." A schematic contour of the step topology was included in each side view to aid the eye. (From Ref. [10], with permission from American Physical Society.)

[70,71]. Without placing undue significance on the negative sign, we focus on the ranking of the formation energies and the corresponding structures. The reconstruction of the flat Si(114) surface consists of rows of dimers (d), rebonded atoms (r), and tetramers (t) in this specific periodic sequence (. . . -d-r-t-d-r-t-d- . . .) along the $[22\bar{1}]$ direction [62]. Since we allowed for a large width of the step region (The size of the step zone was chosen so as to cover one full reconstructed unit cell on each side of the (initially arbitrary) step location), the steps can negotiate their width and location during the genetic evolution. This is apparent in Figure 5.14, which shows that the sequence of motifs (d, r, t) is continued seamlessly from each terrace into the step zone until the atomic structure and the location of the step edge are determined. The best $[\bar{1}10]$-down step structures (Figure 5.14a and b) include a row of hexagons (labeled by "h" in Figure 5.14) in addition to the motifs already encountered on terraces. Other low-energy steps are observed to simply consist of a gap in the -d-r-t- sequence of terrace motifs. For example, Figure 5.14c shows a down-step that contains dimers, rebonded atoms, and tetramers in the correct order, but which are bonded to the upper (lower) terrace via elimination of one tetramer (dimer) row from the -d-r-t- sequence on the terraces. The most favorable up-step structures contain only rows of dimers and rebonded atoms (Figure 5.14d), all motifs in a different order (d-t-r in Figure 5.14e, or only rows of dimers and tetramers in Figure 5.14f).

To provide a closer look at the way the algorithm sorts through different numbers of atoms, we have plotted the lowest formation energy found for every number of atoms N attained *during* the evolution (Figure 5.15a). Such a plot shows that the algorithm can visit, in the same evolution, several structures of particularly low formation energies (magic-number atomic configurations) and select them as part of the genetic pool. Magic-number structures are found for even values of N for both the up and down $[\bar{1}10]$ steps, as seen in Figure 5.15a. The same figure shows that the formation energy of $[22\bar{1}]$-normal steps also has a few distinct local minima, which are located at odd values of N. On the other hand, the $[22\bar{1}]$-shifted configuration does not have local minima that could be identified as magic-number structures. In this case, there are two deep minima for the range of atom numbers spanned, but after the final relaxation they are found to have the same structure only translated by one complete terrace unit cell along the $[\bar{1}10]$ axis.

Finally, in Figure 5.15b, we report the formation energies of the lowest eight structures in the pool for each of the four configurations studied. The figure shows that the steps oriented in the $[\bar{1}10]$ direction have smaller formations energies than those along $[22\bar{1}]$ for all top-ranking structures found. For the $[\bar{1}10]$ direction, the down-steps are easier to form than the up-steps; while for the $[22\bar{1}]$ direction, the up- and down-steps have identical structures and energies. The conclusion that $[\bar{1}10]$ steps are more favorable than $[22\bar{1}]$ steps is consistent with the general expectation that a direction of higher symmetry (i.e., $[\bar{1}10]$) yields lower step energies than a direction of low symmetry.

As seem in Figure 5.14, the algorithm is able to cover a substantial range of step structures that differ in terms of number of atoms, atomic patterns, and location of the step edge. This morphological and structural diversity is a telltale sign of the superior configuration sampling that can be achieved with only one simple genetic operation (the crossover, Figure 5.12b) (In most genetic algorithm implementations, mutations

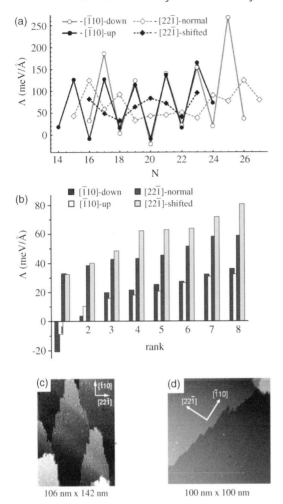

Figure 5.15 Results of the genetic algorithm (a and b) and experimental observations of steps on Si(114) surfaces (c and d). (a) Lowest step energies attained at various atom numbers in the step zone. (b) Final step formation energies of the top-ranking structures for each of the four types of steps described in text. The [$\bar{1}$10] steps have consistently lower energies than similarly ranked [22$\bar{1}$] structures. (c) STM image of a vicinal Si(114) surface obtained after cleaning and brief annealing (reproduced from Ref. [69] with permission from the American Institute of Physics). (d) STM image taken after flashing at 1225 °C followed by 30 min annealing at 450 °C. (Courtesy of D.E. Barlow, A. Laracuente, and L.J. Whitman.) The experiments show that the [$\bar{1}$10] steps are preferentially longer than the [22$\bar{1}$] ones, which is consistent with the calculated step formation energies shown in panel (b). (From Ref. [10], with permission from American Physical Society.)

(small random displacements of arbitrarily selected atoms) are necessary to provide paths for the pool members to evolve toward global minima or toward stationary states. Due to the requirement to allow only structurally distinct pool members at all times, mutations turned out to be unnecessary in the present algorithm).

In the competition between the reduction of the number of dangling bonds in the step zone and the stress created through this reduction, there emerge several structural motifs that are not always mere continuations of the terraces up to the step edge (e.g., the hexagons that appear on the $[\bar{1}10]$-down steps). The presence of hexagons is illustrative of the clear structural asymmetry between the up- and down-steps. The hexagons appear only on the down-steps, while on the lowest-energy $[\bar{1}10]$-up step (Figure 5.14d) we can easily recognize the S_B step [72] formed on the small, dimer-wide (001) nanofacets [62]. As long as the hexagons are present (Figure 5.14a and b), the formation energy of the $[\bar{1}10]$-down steps is clearly lower than that of the similarly ranked up-steps. Interestingly, the discrepancy between the formation energies of the up- and down-steps fades for less favorable $[\bar{1}10]$ step structures (Figure 5.15b), for the reason that those higher-indexed step structures signify a transition from the upper to the lower terrace through intermediate (nano) facets that are formed by dimers and rebonded atoms for *both* the up and down $[\bar{1}10]$ steps. We have observed that the genetic algorithm captures this faceting trend (seen here for the case *of unfavorable steps with high formation energies*); this observation opens up the possibility to study physical systems that are unstable toward faceting within an evolutionary-based global optimization framework.

The formation energies of the steps reported here should in principle be recalculated at the level of electronic structure calculations to refine, as best as possible, their structure and energy ranking. Since these calculations are extremely demanding, we turn toward assessing the validity of our results by comparison with experimental data. We have found that existing experimental observations do support our genetic algorithm results, albeit qualitatively. Laracuente *et al.* [69] reported STM images (reproduced in Figure 5.15c) in which the $[\bar{1}10]$ steps are clearly preferred over the $[22\bar{1}]$ ones even when, due to the preparation conditions [69], the steps may not assume the very lowest-energy structures. This is consistent with the simulation results shown in Figure 5.15b, which indicate that $[\bar{1}10]$ steps have lower energies for *metastable* structures ranked within the first eight at the end of the genetic evolution. More recently, Whitman and coworkers have also imaged step configurations after long anneals at 450 °C. These recent measurements (shown in Figure 5.15d) are more likely to correspond to lowest-energy step structures, and again show that the $[\bar{1}10]$ steps are the longer (and more favorable) ones, thus lending support to our simulation results. In terms of atomic positions, so far we have found no published data on the structure of steps on Si(114). We have proposed here several low-energy structures (Figure 5.14), which are amenable to experimental testing via high-resolution STM measurements combined with *ab initio* density functional calculations.

5.2
Genetic Algorithm for Interface Structures

Grain boundaries (GBs) play an important role in microstructure stability, mechanical behavior, and transport properties of many polycrystalline materials. Grain boundaries are major defects affecting the performance of many microelectronic materials and devices, such as micromechanical materials, nanocrystalline materials, and solar energy application devices [73–76]. Therefore, understanding the structures of grain boundaries at the atomic level is highly desirable. However, structural complexity of grain boundaries makes both experimental and theoretical studies difficult. Although the experimental tools such as HRTEM are widely used in studying grain boundaries in materials, the experimental resolution necessary to examine the detailed atomic structure of grain boundary is still difficult to achieve. On the other hand, many possible structures for a given grain boundary also make theoretical study complicated [76–82]. The relatively large system size (hundreds to thousands atoms) makes first-principles calculations very costly. Meanwhile, many classical potentials, which are fast and useful for calculating systems with large size, need to be checked for their accuracy. In this section, we will analyze an efficient GA procedure to generate grain boundary structures.

5.2.1
GA for Grain Boundary Structure Optimization

Symmetrical tilted Si[001] GB can be constructed by rotating two Si grains in the opposite directions by the same angle around the [001] rotation axis and then adjoining the two grains together [83]. Thus, in the direction of boundary plane the GB will show periodic structures. We model the GB using a simulation block with periodic conditions in the directions parallel to the boundary plane (x and y). In the perpendicular direction z, the grains are terminated at free surfaces. The geometry and dimensions of the computational cells are shown in Figure 5.16. Atomic positions in the cell are generated from geometric coincident site lattice construction. In this geometry, the rigid-body translations are free to occur if they lead to more energetically favorable structures. The thickness of the grains in the z-direction is large enough to exclude long-range elastic interactions between the GB and surfaces. This is ensured by performing calculations with different thicknesses and

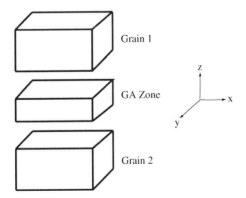

Figure 5.16 The computation cell for the GA search. The x- and y-directions are periodic, while z-direction is terminated at free surfaces. The interface area between the two grains is called the GA zone. (From Ref. [9], with permission from American Institute of Physics.)

verifying that the results are nearly the same. In this chapter, the thickness is chosen to be 10 times of the period in the x-direction.

GA is used to generate and optimize candidate structures for the grain boundary. The atomic structure of grain boundaries can be seen as a reconstruction of the interface atoms between two grains. We first define a GA zone by choosing a slab at the interface between the two grains, as shown in Figure 5.16. As in Section 5.1.4, the GA operations will be executed on the atoms inside the GA zone, while atoms outside the GA zone are allowed to relax to a local minimum of the potential energy after each GA operation; most details of the implementations are similar to those in Refs [2,10].

5.2.2
Structures Generated by GA

A symmetrical Si[001] tilt grain boundary can be described as a GB with median plane (110) or (100) and a rotation angle in the range 0–90° [81]. We searched for the structures of Si GBs at 18 disorientation angles with GA using the classical Stillinger–Weber potential. Each pool contains the ground-state structure, which is the structure with the lowest energy calculated using the Stillinger–Weber potential, together with many higher-energy metastable structures. These lowest-energy structures are in agreement with previous theoretical and experimental results while they are produced by GA with much higher efficiency. In Figures 5.18–5.21, we show the GB structures of four different orientation angles from 0 to 90°. In these figures, we show not only the lowest-energy structures but also some of the metastable structures. After a careful look at the results, we can see that the structural unit model is generally a good description of the structures we have found. The structural unit model describes GBs in terms of structural units consisting of either the perfect-crystal structure or dislocation cores with the associated Burgers vectors [82,84]. The majority of Si[001] symmetrical tilted GBs in the entire range of disorientation can be constructed from three characteristic structural units as shown in Figure 5.17:

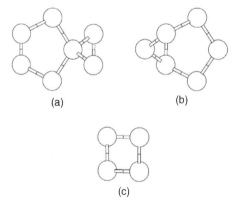

Figure 5.17 The structural units for the Si[001] symmetrical tilted grain boundaries: A is a core of pure edge dislocation, B is a mixed core dislocation with screw component, and C is the unit of a perfect-crystal structure. (From Ref. [9], with permission from American Physical Society.)

i) Unit A, a pure edge dislocation core, with a Burgers vector $\mathbf{b} = (1/2)a_0\,[110]$, where a_0 is the lattice constant of the perfect crystal.
ii) Unit B, the core of a 45° mixed dislocation, with the Burgers vector inclined at 45° to the rotation axis. This mixed dislocation core has a screw component, which is parallel to the rotation axis.
iii) Unit C, the unit of a perfect crystal.

Also we can see that there are generally two basic ways how these structural units are connected to form the grain boundary: the straight way or the zigzag way. The sequence of the structural units arranged and the way they are connected will have an impact on the energies of the GBs, which we will illustrate in the following examples.

In Figure 5.18, we show a typical example of the small-angle grain boundary with tilted angle 16.26°. For ground-state structures with small angles, the grain boundary structures can often be seen as rows of parallel pure edge dislocations (unit A), which is a pentagonal–triangular pattern sharing the same edge, separated by a series of good crystal units (unit C). In this case, the structure Figure 5.18a has two dislocation core structures separated by five crystal structure units between them, which gives a CCCCCACCCCCA arrangement in the GB area. This structure also has a mirror-like symmetry across the boundary plane. The structures shown in Figure 5.18b–d are some of the higher-energy structures in the pool. Among these, only the structure in Figure 5.18c preserves the mirror-like symmetry across the GB

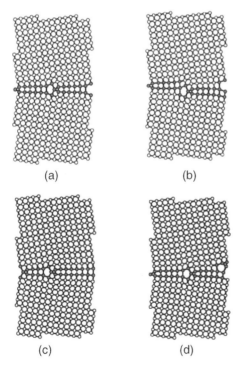

Figure 5.18 Atomic structures of $\Sigma = 25$ Si[001](340) GB with $\theta = 16.26°$. (From Ref. [9], with permission from American Physical Society.)

plane. Structure in Figure 5.18c has the same set of units as the structure in Figure 5.18a. However, it has a different unit arrangement, CCCACCCCCCA, which results in higher GB energy. Structure in Figure 5.18b has similar units arrangement, but the dislocation cores are shifted apart in the [110] direction, which causes a higher energy. Structure in Figure 5.18d can be seen as formed by the addition of extra dislocations to the structure in Figure 5.18a, which will also increase the GB energy. Besides these four structures, there are also numerous new structures in this rotation angle in the GA pool, which shows the multiplicity of GB structures in a given rotation angle. This multiplicity, which demonstrates the complexity of GB material, is also found in experiments [85,86].

It is not surprising that the small-angle GBs have many units of perfect-crystal lattice as the small angle will add only a small perturbation to the perfect-crystal-like structure and introduce some dislocation cores [81]. The majority of the crystal-like units would like to be preserved. However, when the rotation angle becomes bigger, the number of crystal units shrink and more dislocation core structures are introduced. In Figure 5.19, the rotation angle is increased to 36.87°. Structure in Figure 5.19a has a structure unit arrangement as ACAC in one period, which has two unit A dislocation cores separated by a single-crystal unit C. Meanwhile, the

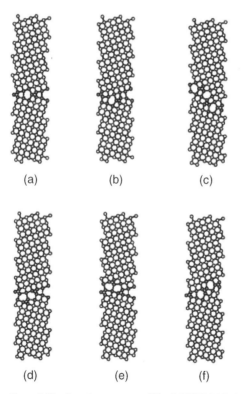

Figure 5.19 Atomic structures of $\Sigma = 5$ Si[001](120) GB with $\theta = 36.87°$. (From Ref. [9], with permission from American Physical Society.)

periodicity in the grain boundary plane becomes much shorter compared to the low-angle grain boundary structures in Figure 5.18. In Figure 5.19, other structures have more complicated arrangement of dislocation cores and form zigzag boundaries.

In Figure 5.20 when the rotation angle is further increased to 53.15°, the (130) boundary is formed and the period in the boundary plane is even shorter. Figure 5.20b has two dislocation core units A connected together and crystal unit is absent, making the structure units arrangement simply AA. What is interesting is that this is not a lower-energy structure compared to that in Figure 5.20a, which has a zigzag structure of two A units. We can find that other metastable structures in Figure 5.20c–f have more complicated dislocations connected together and structure in Figure 5.20e cannot be described by the three basic structural units A, B, and C defined above as the triangular and pentagonal rings are separated by a crystal unit. This might suggest that when the rotation angle becomes big, the matching of the two grains becomes flexible compared to the low-angle case and more exotic structures may appear, which is further illustrated in Figure 5.21.

In Figure 5.21, the rotation angle is 67.38°, making the GB a high-angle one. Structures in this rotation angle have been studied extensively both experimentally and theoretically in the literature [83,86,87]. A variety of structures exist in this angle

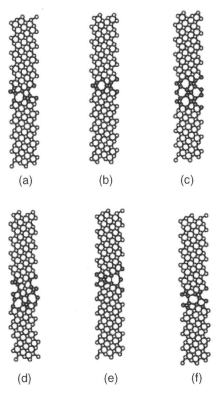

Figure 5.20 Atomic structures of $\Sigma = 5$ Si[001](130) GB with $\theta = 53.15°$. (From Ref. [9], with permission from American Physical Society.)

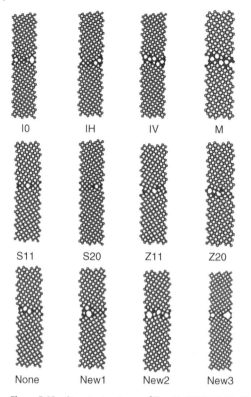

Figure 5.21 Atomic structures of $\Sigma = 25$ Si[001](340) GB with $\theta = 67.38°$. (From Ref. [9], with permission from American Physical Society.)

and interestingly, the structure with the lowest energy is the one that has the most dislocation dipole content [83]. Using GA, we have successfully reproduced all the structures reported in the literature and we also got some new structures constructed with different units. The structures in Figure 5.21 are named according to the literature [83]. From the structural unit viewpoint, most of the (150) structures can be constructed using a mixed dislocation core, B with crystal units C. For example, (S20) and (S11) in Figure 5.21 have straight arrangements, BBCC and BCBC, respectively, while (Z20) and (Z11) in Figure 5.21 have zigzag arrangements, BBCC and BCBC, respectively. There are also exceptions that the atomic structures cannot be explained by the basic units arrangement. In Figure 5.21, (I0), (New1), and (New2) all have very rare six-member rings that are not found in GBs of other rotation angles. In Figure 5.21, (New3) are similar to (I0) but with a larger outward shift between the two grains, which makes an eight-member ring structure from the six-member ring structure. The multiplicity of these structures might be attributed to the fact that high-angle grain boundary gives more flexibility to atoms reconstructing in the GB area. Overall this high-angle grain boundary (150) has more dislocation cores than the low-angle ones. These structures have similar energies and that may be the reason why multiple structures can be observed in this angle experimentally with HRTEM and Z-contrast electron technique [87].

5.2.3
Grain Boundary Energy Calculations

The structures obtained from the GA search are then used as starting points for first-principles and tight-binding calculations of the GB energies. This is done in the following way: a fully periodic system can be made by matching the two identical boundaries in the z-direction together so that it will have periodicity in direction perpendicular to the boundary plane in addition to the periodicity in the x–y plane from the GA-generated structure. This method is similar to the one used in Ref. [83]. In the new structure, the two grain boundaries that have opposite directions are approximately 50 Å apart, depending upon the details of the structures. The number of atoms ranges from 200 to over 1000 for all the structures we constructed. In these calculations, the dimension normal to the grain boundary is roughly 10 times larger than the dimension along the boundaries, which are similar to the surface slab calculations. A set of four special k points is chosen to sample the two-dimensional rectangular Brillouin zone. All the atoms in the system are allowed to relax until the forces are less than 10 meV/Å The first-principles calculations are performed using the VASP code and tight-binding calculations are performed using the environment-dependent Si tight-binding potential developed by Wang and coworkers [88]. The grain boundary energy is defined as

$$\gamma_{GB} = \frac{E_{slab} - N\mu}{A_{GB}}, \tag{5.2}$$

where n is the number of atoms in the structure, μ is the bulk energy of silicon, and A_{GB} is the grain boundary area.

Since first-principles calculations require high computational resources for large systems, a number of selected systems with shorter periods along the grain boundaries, Si[001](120) and Si[001](130), are computed using *ab initio* techniques here. These are also the relatively high-angle grain boundary structures that can exhibit multiple structures with similar energy. We also used environment-dependent tight-binding potential model for Si to test how well it produced the grain boundary energy [88]. This potential goes beyond the two-center approximation by allowing the hopping terms to be modified by local atomic environment, which yields an improved transferability away from the bulk environment.

In Figure 5.22, we show the GB energies of several Si[001](120) and Si[001](130) structures. These energy values are obtained from *ab initio* calculations, environment-dependent tight-binding potential, and classical Stillinger–Weber potential. One striking observation is that result using tight-binding potential agrees very well with the first-principles calculations, while the classical potential gives significant deviations. In particular, the tight-binding result gives the correct energy ordering and reasonably good energy difference in the structures, while the classical potential cannot. This suggests that the classical potential, which performs fast in calculation, is good to be used in the genetic algorithm search to explore all the possible structures with high efficiency. However, the TB model can be effectively used as a calculation tool for examining these grain boundary energies even with large unit cells. It has also

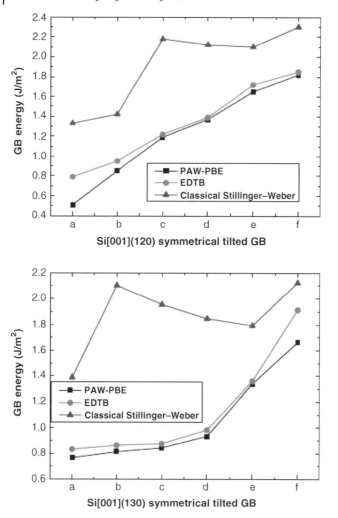

Figure 5.22 GB energies for Si[001](120) and Si[001](130) calculated via DFT, environment-dependent tight-binding potential, and Stillinger–Weber potential. (From Ref. [9], with permission from American Physical Society.)

been shown previously that the environment-dependent Si tight-binding potential also gives a good description of the energies Si[100](150) grain boundary [83].

In Figure 5.23, we show the lowest energies of GB in a given angle as a function of the total angles from both classical potential and tight-binding calculations. It can be seen that some lower-period structures such as Si[100](120) and Si[100](130) have energies as the local minima among the surrounding structures. Thus, these structures, which have special angles and low periodicity, usually have the higher possibility of being observed in the experiment. In fact, faceting of the grain boundary can usually happen when the rotation angle of the GB is close to these special angles. Then faceting provides an efficient way of reducing the GB energies.

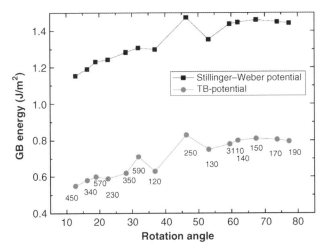

Figure 5.23 Grain boundary energy as a function of rotation angle. (From Ref. [9], with permission from American Institute of Physics.)

In summary, GA is highly efficient and accurate for the grain boundary structure generation and reproduces all the structures observed in experiments as well as deduced by the theoretical calculations. Almost all GB structures can be expressed in the structural unit model except for a few high-angle cases in which six- and eight-member ring structures appear. Starting from the low-angle grain boundary, as the disorientation angle between the grains becomes bigger, the number of perfect-crystal unit in the grain boundary area will decrease and the connection between dislocation cores will become complex. The environment-dependent tight-binding potential for Si is used to evaluate the energy of selected grain boundary structures and have a very good agreement with the first-principles calculation results. It provides us a useful tool for evaluating the energy orders of the structures and making corrections for the traditional classical potential results. Besides the symmetrical tilted GB, more types of GBs, including the twisted and asymmetrical ones, can also be explored by the genetic algorithm in the similar way. This would be done in the later work. For Si GBs, the environment-dependent tight-binding potential would always be a valuable tool for the GB energy calculation due to its high transferability.

5.3
Nanowire and Nanotube Structures via GA Optimization

5.3.1
Passivated Silicon Nanowires

The continuous miniaturization of electronics industry has achieved the limit in which the interconnection of the devices in a reliable and controllable way is particularly challenging. Fervent strides are underway in the preparation of nanoscale wires for molecular and nanoelectronics applications [89]: such wires (possibly doped or functionalized) can operate as both nanoscale devices and interconnects [90]. Silicon nanowires (SiNWs) offer, in addition to their appeal as building blocks for nanoscale electronics, the benefit of simple fabrication techniques compatible with the currently well-developed silicon technology.

The current growth methods [91–94] can yield wires with diameters ranging from several tens of nanometers down to 1 nm. These SiNWs are usually crystalline with only a few axis orientations observed and have a prismatic shape bounded by facets that are parallel to the wire axis [92–96]. While remarkable progress has been achieved in terms of preparation and characterization of SiNWs, atomic-level knowledge of the structure remains necessary for a complete understanding of device properties of these wires. At present, attempts to predict the structure of SiNWs are affected by the lack of robust methodologies (i.e., algorithms coupled with model interactions) for searching the configuration space, and most studies to date rely on heuristically proposed structures as starting point for stability studies of SiNWs at the *ab initio* level. Electronic structure calculations are too computationally demanding to be used in a thorough sampling of the potential energy surface of SiNWs. On the other hand, most of the empirical potentials for Si are fast, but are not sufficiently transferable to capture accurately the structure and energetics of various wire configurations. Despite these obstacles, the current theoretical efforts to find the structure of SiNWs have been very vigorous and have led to the identification of a number of low-energy configurations for pristine SiNWs [97–101]. In comparison, the structure of passivated nanowires has received much less attention from a theoretical perspective, although most of the experimental techniques to date yield wires that are passivated either with oxides [102,103] or with hydrogen [94,95].

Motivated by the STM experiments of Ma *et al.* [94], we have investigated the structure of thin H-passivated nanowires oriented along the [110] direction. To this end, we have designed a robust and efficient optimization procedure based on a GA, followed by structural refinements at the density functional theory level. We have found that in the presence of hydrogen, the silicon atoms of the nanowire can maintain their bulk-like bonding environment down to subnanometer wire dimensions. Furthermore, our calculations reveal that as the number of atoms per length is increased, there emerge three distinct types of wire configurations with low formation energies (magic wires). Two of these structures have a plate-like aspect in cross section, which have not been observed so far. The third one has a hexagonal section, which is consistent with recent experiments for Si and Ge wires [94,104]. Given their extremely small diameters, the magic structures found here by the combined GA–DFT optimization procedure cannot be predicted by thermodynamic considerations based on the Wulff construction [105]. The procedure described below is generally applicable for finding the structure of any ultrathin nanowire provided that suitable models for the atomic interactions are available and that the spatial periodicity along the wire axis is known.

The choice of GA for the present work was motivated by previous findings that search procedures inspired by the genetic evolution can successfully be used to determine the structure of Si clusters [106] and surfaces [2]. The SiNWs are modeled using a supercell that is periodic in one dimension, with the period set according to experiments [94]. We choose the Hansel–Vogel (HV) potential to describe the atomic interactions, for this model has been shown to reproduce well the energies of hydrogenated phases of the Si(001) surface [107].

The algorithm for SiNW optimization is similar to that we designed for high-index Si surfaces, so we focus here on the departures from Ref. [2]. During a GA optimization run, a pool of at least 60 structures (which are initially just random collections of atoms) is evolved by performing mating operations; such operations consist in selecting 2 random structures (parents) from the pool, cutting them with the same plane parallel to the wire axis, and then combining parts of the parent structures that lie on the opposite sides on the cutting plane to create a new structure (child). The child structure is first passivated by satisfying all its dangling bonds with H atoms and then relaxed with the HV model [107]. We include the child structure in the genetic pool based on its formation energy f defined as

$$f = (E - \mu_H n_H)/n - \mu, \tag{5.3}$$

where E is the total energy of the computational cell with n silicon atoms (n is kept fixed during a GA run) and n_H hydrogen atoms, μ is the (reference) bulk cohesive energy of Si in its diamond structure, and μ_H is the chemical potential of hydrogen. The H chemical potential is set such that certain hydrogenation reactions at surfaces are thermodynamically possible. These reactions are shown in Figure 5.24a, along with our two chosen values for μ_H. The pool is divided into two equal subsets, one for each values of μ_H. The mating operations are performed both with parents in the same subset and with parents in different subsets, in order to ensure a superior sampling of the potential energy landscape. The mating is carried out 15 times during a generation, and a typical GA run has 50 000 generations. At the end of each run, all structures are relaxed using the VASP package [108] (The *ab initio* calculations are performed within the generalized gradient approximation [109]. The kinetic energy cutoff is set at 11 Ry and the Brillouin zone is sampled using 16 k-points. The SiNW is positioned at the center of a supercell with a vacuum space of 12 Å separating the periodic images of the wires. Each SiNW structure is relaxed until the magnitude of the force an atom is smaller than 0.01 eV/Å). The chemical potential μ_H used to compute DFT formation energies is determined so that it maximizes the correlation with the HV formation energies for a few hundred configurations (refer to Figure 5.24b). Therefore, while some energetic reordering does occur after the DFT calculations, most of the low-energy structures found with the HV model remain relevant at the DFT level, especially in the range of DFT formation energies below 0.15 eV per atom.

The most stable structures (DFT level) that we found through the above procedure using $\mu_H = -3.45$ eV are shown in Figure 5.25 for different numbers of Si atoms in the range $10 \leq n \leq 30$. The second half of the genetic pool [corresponding to $\mu_H = -3.35$ eV (DFT)] retrieved the same structures for all even values of n. In the case of odd n, the two sides of the pool found structures that are different only in the position of one peripheral Si atom. As Figure 5.25 reveals, the GA–DFT optimization finds stable structures that are made of six-atom rings (viewed along the SiNW axis), with some of the rings incomplete depending on the value of n. The six-atom rings

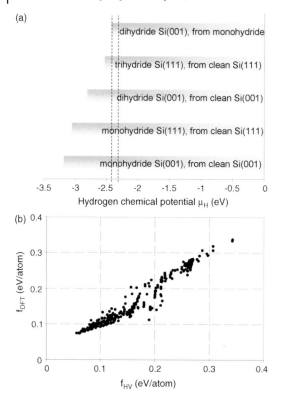

Figure 5.24 (a) Choice of the chemical potential μ_H used in the genetic algorithm search with the HV potential. The vertical lines are located at $\mu_H = -2.42$ eV and $\mu_H = -2.32$ eV. Each horizontal bar shows the μ_H range for which a given hydrogenation reaction is favorable. (b) Formation energy f_{DFT} at the DFT level [108] versus the same quantity computed with the HV model [107] at $\mu_H = -2.42$ eV. The chemical potential $\mu_H = -3.45$ eV for the DFT calculations was determined such as to maximize the correlation between the DFT and HV formation energies for 422 structures (data points). The magic wires described in the chapter lie below 0.15 eV, where the agreement between the DFT and the HV values is best. (From Ref. [11], with permission from American Chemical Society.)

can arrange themselves in chains (e.g., $n = 12$, $n = 16$ in Figure 5.25); for $n > 16$, the rings form fused pairs of such chains ($n = 18, 24, 30$), which we call double-chains. Structures similar to the double-chains shown in Figure 5.25 have also been considered in recent studies of quantum confinement in SiNWs [110]. The formation energy of the most stable structures for $10 \leq n \leq 30$ retrieved is plotted in Figure 5.26 as a function of n. When the value of n increases, the formation energy assumes an overall decreasing trend while displaying odd–even oscillations. The relatively larger formation energy of the odd-n structures corresponds to configurations where one Si atom protrudes from the surface of the wires, thus creating a somewhat unfavorable bonding environment.

As shown in Figure 5.25, the ground state for each n has a bulk-like structure with no significant reconstructions. The absence of reconstruction is due to the hydrogen environment at chemical potentials μ_H that favor no less than five hydrogenation

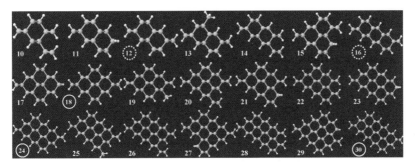

Figure 5.25 Nanowire structures (axial view) with $n = 10$ to $n = 30$ silicon atoms per unit cell, found after genetic algorithm optimizations and the subsequent DFT relaxations. The H atoms are the smaller white spheres. The values n that correspond to magic wires are indicated by dashed and solid circles for chain and double-chain SiNWs, respectively. (From Ref. [11], with permission from American Chemical Society.)

reactions (refer to Figure 5.24). The use of smaller μ_H values can lead to understanding the interplay between reconstruction and hydrogen coverage on the SiNW facets, an investigation that we will address in subsequent studies. For smaller values of n ($n < 17$), the common motif present in lowest-energy structures is the six-atom ring. When chains of complete six-atom rings are formed, the structures become particularly stable, as it is in the case of $n = 12$ and $n = 16$ SiNWs, which are made of $R = 2$ and $R = 3$ complete rings, respectively. The six-atom ring chains expose two $\{111\}$ facets with monohydride terminations. Another common feature of all (complete) chain structures is the presence of two dihydrides with the SiH_2 planes oriented perpendicular to the wire axis. For $n > 17$, the most favorable structures are the double-chains described above (and illustrated in Figure 5.25), whose building blocks we call double-rings. These blocks are readily identifiable in the case of $n = 18, 24,$ and 30, which correspond to a number of $D = 2, 3,$ and 4 full double-rings, respectively. The double-chains expose a total of four $\{111\}$ nanofacets

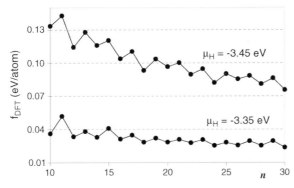

Figure 5.26 DFT formation energies per Si atom for two values of the H chemical potential, plotted as a function of n. Structures with even n are local minima at *both* the empirical [107] and *ab initio* [108] levels. (From Ref. [11], with permission from American Chemical Society.)

that form parallelogram-shaped SiNW cross sections. The complete double-chains also contain two dihydrides, identical to those capping the chain SiNWs.

We have conducted further optimization studies for $n > 30$ using, however, only even atom numbers n. We found that starting at $n = 60$, structures with hexagonal cross section become stable over the double-chain SiNWs described above. Even the smallest hexagonal SiNW ($n = 28$) has a formation energy that is only slightly higher (by 0.005 eV/atom) than that of the double-chain with the nearest size ($n = 30$). The hexagonal wires are bounded by two {001} and four {111} facets, and can be described by their number of completed concentric layers L of six-atom rings. Due to the relative stability and structural closure of the chain, double-chain, and hexagonal SiNWs, we name these configurations *magic*. The shapes of prototype magic SiNWs are illustrated in Figure 5.27, and a description of their building blocks and numbers of atoms is summarized in Table 5.3. The formation energies of magic SiNWs were

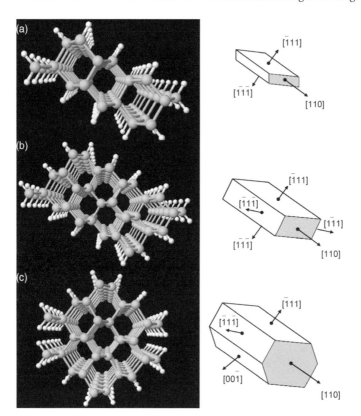

Figure 5.27 Magic nanowires (perspective view) found as minima of the formation energy per atom. The chain (a) and double-chain (b) are characterized by the number of complete six-atom rings R and double-rings D, respectively (also refer to Table 5.3). The configurations with hexagonal cross section have a number L of full concentric layers ($L = 2$ in panel (c)) of six-rings and are consistent with recent observations of H-passivated SiNWs [94,95]. The facet orientations of magic wires are shown on the right. (From Ref. [11], with permission from American Chemical Society.)

5.3 Nanowire and Nanotube Structures via GA Optimization

Table 5.3 Building blocks and magic numbers (n) of Si atoms corresponding to the different types of low-energy SiNWs found in the global optimization.

Structure	Building units	n	Mono-H	Di-H
Chain	Rings, R	$4(R+1)$	$2(R+1)$	2
Double-chain	Double-rings, D	$6(D+1)$	$2(D+2)$	2
Hexagon	Layers, L	$2L(3L+1)$	$4L$	$2L$

The last two columns show the number of mono- and dihydrides in each case.

separately plotted as a function of n in Figure 5.28 at both levels of theory [107,108]. The plot reflects the structural trends described above, namely, the transition between chain and double-chain at $n = 16$ and the transition between double-chain and hexagonal SiNW that starts at $n \approx 29$. It should be noted that *incomplete* hexagons appear frequently in the range $28 < n < 60$; therefore, the latter shape transition is not as well defined as the former.

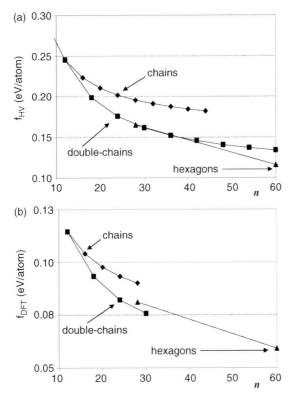

Figure 5.28 Formation energies per Si atom computed for specific structures of the H-passivated [98] nanowires: chain (diamonds), double-chain (squares), and hexagonal (triangles), computed at the level of HV potential [107] with $\mu_H = -2.42$ eV (a) and of DFT calculations [108] with $\mu_H = -3.45$ eV (b). (From Ref. [11], with permission from American Chemical Society.)

We note that even in the absence of reconstruction, the shape of SiNWs (refer to Figure 5.27) departs markedly from the equilibrium crystal shape predicted by the constrained minimization [105] of the overall surface energy. The reason for this departure is that the number of surface Si atoms is larger than, or comparable to, the number of bulk-like atoms (refer to Table 5.3). For the chain and double-chain SiNWs, the polygonal shape itself is not preserved as only two or four facets are exposed instead of the expected six [94,95]. For the ultrathin hexagonal wires, the departure from the Wulff construction [105] is more subtle. Although the cross section is hexagonal, the relative size of the $\{100\}$ and $\{111\}$ facets is not predicted by the surface energies because the interactions of facet edges are significant for wires thinner than 2 nm.

To our knowledge, chain and double-chain wires have not been observed so far for SiNWs, perhaps because the diameters achieved in laboratory are larger than what is favorable for plate-like structures to form [94]. Recent experiments were successful in isolating and characterizing [98] SiNWs with hexagonal section and diameters between 1.3 and 7 nm [94]. The thinnest hexagonal wire in our work ($L = 2$ in Table 5.3) has a diameter of approximately 1.2 nm, as estimated by the distance between its $\{001\}$ facets. The first hexagonal wire that we find to be more stable than any double-chain magic wire is about 1.8 nm in diameter, that is, already in the range of the experimentally reported hexagonal SiNWs [94]. Pursuing further the comparison between the hexagonal wires found here and those in Ref. [94], we have computed STM images for the facets of the $L = 5$ hexagon. The simulated images are shown in Figure 5.29 a for a $\{111\}$ facet covered with monohydride, and for a $\{001\}$ facet covered with dihydride. Our calculations are in agreement with the STM experiments, which also showed the exclusive presence of dihydride species on the $\{001\}$ facets of [110] SiNWs. Furthermore, the resemblance (Figure 5.29) between the simulated STM image and the experimental one brings strong support to the predictive power of the theoretical methodology presented here.

In conclusion, we described a combined GA–DFT procedure to search for the structure of [110] passivated SiNWs, and presented our results for [110] nanowires with up to 60 atoms per unit cell. Subject to the availability of reasonable empirical potentials, this GA–DFT procedure can be adapted quite readily for finding the structure of any type of nanowire that exhibits atomic-scale periodicity along its axis. For H-passivated [98] silicon wires, the genetic search revealed three types of magic structures (shown in Figure 5.27): chain, double-chain, and hexagonal SiNWs. Our results for hexagonal wires are consistent with recent experiments [94]. Given this agreement with experiments at diameters larger than 1.8 nm, it is conceivable that the chain and double-chain structures proposed here can be observed experimentally upon pursuing the preparation of SiNWs thinner than 1.3 nm.

5.3.2
One-Dimensional Nanostructures under Radial Confinement

In this section, we present another application of genetic algorithms: we show that a real-space, variable-number algorithm can be used to retrieve simultaneously *the lowest-energy structure and the optimal number of atoms* of one-dimensional

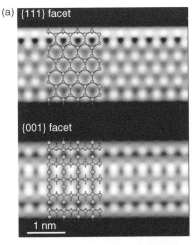
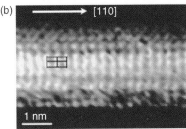

Figure 5.29 (a) Simulated empty-state STM images (bias of +2.0 V) of {111} and {001} facets of an $L = 5$ hexagonal nanowire. (b) Actual STM image of a {001} facet. (From Ref. [11], with permission from American Chemical Society.)

nanostructures subjected to desired conditions of radial confinement, starting from a single atom in the periodic unit cell. This algorithm is based on two-parent crossover operations and zero-penalty "mutations," the latter allowing the algorithm to evolve even from a genetic pool made of identical structures. We show that a sufficiently rich set of crossover operations (attempted with equal probability) can make the procedure effective for finding the atomic structure very different 1D nanomaterials. We test the algorithm for carbon nanotubes (CNTs) and for Lennard-Jones (LJ) nanotubes using runs where crossovers are applied either individually or in combination. By analyzing the acceptance probabilities of the structures created in these runs, we discuss the algorithm performance and possibilities for improvement.

5.3.2.1 Introduction

One-dimensional nanostructures presently show tremendous technological promise due to their novel and potentially useful material properties. The continuous miniaturization of electronics industry has achieved the limit in which the interconnection of devices in a reliable way is particularly challenging. Efforts are underway for synthesizing nanowires for specific molecular and nanoelectronics applications. Such wires, possibly doped or functionalized, can operate as both nanoscale devices and interconnects [90]. While remarkable progress has been achieved in terms of preparation and characterization of new one-dimensional materials [91,94–96,111], atomic-level knowledge of the structure remains necessary for a complete understanding of usefulness of these nanowires for the device applications. Predictions of the structure of nanowires may be at present affected by the lack of robust methodologies (i.e., search algorithms coupled with model interactions) for searching the configuration space, and most studies to date rely on heuristically proposed structures as starting point for further stability studies at the *ab initio* level [97–101]. Recently however, genetic algorithms coupled with empirical interaction potentials have made their way into the field of predicting nanowire and nanotube structures [112–115].

Most developments of genetic algorithms have occurred in area of structure optimization *for clusters*; these developments include, for example, extended compact genetic algorithms [116], differential evolution [117], and particle swarm algorithms [117,118]. As we have seen in the previous sections, recent extensions of genetic algorithms to 2D and 3D periodic systems have proven very versatile for finding surface reconstructions and predicting crystal structures and polymorphs [3]. The key ingredient for the *efficient* use of GAs in the global optimization and structure prediction for 2D and 3D systems is the provision that the number of atoms is allowed to vary. Such provision, which is necessarily absent in the case of clusters, facilitates optimization pathways than span systems with different numbers of atoms, thus providing fast routes toward the global minimum of the appropriate fitness function. To our knowledge, the use of *variable-number genetic algorithms* for the global optimization of 1D nanostructure has not been reported so far, as the applications of GAs for 1D systems used constant numbers of atoms in the periodic cell [112–115].

Here we present a novel application of genetic algorithms, namely, their use for finding the structure of a 1D nanotube via *simulated growth*. Because the simulation of growth is envisioned, we choose a *variable-number* genetic algorithm, which in the case of 2D and 3D periodic structures was shown to retrieve the correct global minimum of the relevant energetic quantity – that is, surface energy in 2D and cohesion energy per particle in 3D. The 1D nanotubes investigated here are started with very few atoms in the periodic cell. The growth of the nanostructures takes off and proceeds *solely* through crossover operations and stops when the optimal structure (i.e., that with lowest energy per particle) for the given confinement conditions is found. As such, the growth is not a reflection of the kinetic processes that occur in actual synthesis experiments, but it is rather a different way to seek the *optimal* nanostructure that can be synthesized under the prescribed confinement conditions.

5.3.2.2 Description of the Algorithm

Nanotubes are simulated using periodic boundary conditions along their axis [119] and spatial confinement in the radial direction. The spatial period along the nanotube axis is kept constant, and the radial confinement is achieved by placing hard repulsive walls at two desired values of the radius [120]. The nanotubes will be formed between these two cylindrical walls, whose sole purpose is to keep the system from expanding outside a desired cylindrical annulus. Our test systems are carbon atoms modeled via the Tersoff potential [121], and Lennard-Jones particles [122] with the parameters $\sigma = 1.5$ Å and $\varepsilon = 2.0$ eV. To increase the speed of the calculations, the range of the LJ potentials is truncated at $r_c = 2.7$ Å; this value has proved to be sufficiently large, such that the structures retrieved by our GA runs do not change if the cutoff is increased further. Next, we give a description of the algorithm that focuses on the genetic operations used. For the systems considered here, "generation zero" (the starting set of structures) consists in a genetic pool of $p = 40$ configurations, each member having a small number of atoms placed at random locations inside a cylindrical annulus.

Selection. The selection is based on a single fitness function, namely, *the potential energy per particle*. A new structure (i.e., the result of a crossover operation) is included in the genetic pool if its energy is smaller than the highest energy among the structures already present *and* if it is no closer than $\delta = 0.001$ eV to the energy of any one of the pool structures that has the same number of atoms. This means that a new structure is allowed in the pool despite having an energy close to that of another structure, *provided* that it has a different number of atoms. Allowing structures with same energy but different number of atoms to enter the genetic pool helps to get the growth process started when the initial population members have one or very few atoms. Indeed, when particles are far apart and do not interact, allowing a child with a larger number of atoms (that are still far enough that interactions are negligible) will increase the particle density and help the growth of the structure. The selection process that we use prevents the duplication of members in the pool (whose size is kept at $p = 40$), duplication that often keeps the algorithm from retrieving the global minimum of the fitness function. For this GA procedure, we impose no restriction on the number of atoms of any newly created structure, with the obvious exception that a structure with zero atoms is not allowed; the reason for leaving the number of atoms variable is that we want the optimal number of atoms to be found at the same time as the optimal atomic structure. Admittedly, we do not know beforehand that the number of atoms would actually converge. The choice for a variable-number GA is only justified *a posteriori*, by the results obtained.

Crossovers. A crossover (which we often also call a *move*, for simplicity) is an operation carried out by splicing two randomly chosen (parent) structures from the genetic pool in order to create a new (child) structure. The new structure is considered for inclusion in the pool as explained in the paragraph above. There are two types of crossover operations that we employ in this study. The first type, which we call *sine* crossovers, requires cutting the parents along sinusoidal lines that are compatible with the periodic boundary conditions along the axis of the tube and with respect to the angular coordinate. This procedure is adapted from recent work in 3D crystal structure prediction [3] that showed that at least for 3D

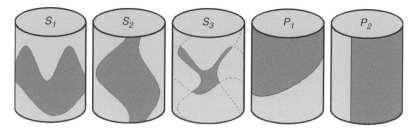

Figure 5.30 The crossovers S_1, S_2, S_3, P_1, and P_2 used in the genetic algorithm optimization of 1D tubular nanostructures. The different shades on the cylindrical surface indicate the domains that are taken from each parent to create a new structure. (From Ref. [13], with permission from Taylor and Francis.)

periodic systems, the real-space GA is more efficient when using cutting functions that obey the periodic boundary conditions.

We have used three specific sine crossovers. In these crossovers, sine functions of randomly chosen frequency, amplitude, and initial phase were used to create two (e.g., S_1 or S_2 in Figure 5.30) or four domains (S_3 in Figure 5.30) on each of the two parent structures. In the case of S_1 and S_2, we assemble one domain from each parent to create the child. For S_1, there are two cutting functions, each of which is a sine function that closes on itself and has the polar angle ϕ as variable. Similarly, in the case of S_2, there are two cutting functions that depend on the axial coordinate z and obey the periodic boundary condition along the z-axis. The move S_3 uses four sine functions in order to create four domains on each parent (Figure 5.30). The child structure created with the S_3 move combines three domains on one of the parents with one domain of the other parent. Given that the contribution of one parent is larger in the final structure, for certain values of the amplitudes, phases, or frequencies, the single domain that comes from the second parent can be viewed merely as a perturbation on the configuration of the first parent. Therefore, while we *do not have explicit mutations* in the algorithm, the S_3 move can act like a mutation of the first parent when the domain that comes from the second parent (darker shade in the S_3 viewgraph of Figure 5.30) is very small.

The second type of crossovers that we used is based on planar cuts as in the pioneering work of Deaven and Ho. We considered cuts with planes of arbitrary orientations, which are obviously not compatible with the periodic boundary conditions (refer to panel P_1 in Figure 5.30, in which the upper and lower part belong to different parent structures and thus create a parent-domain boundary and the z-bounds of the supercell). We also use, separately, crossovers based on planes parallel to the symmetry axis of the system, which are compatible with the boundary conditions (P_2 in Figure 5.30). The sine and planar crossover types were used individually in GA runs, as well as in combinations. The combination we describe here is the one in which all moves in Figure 5.30 are attempted with equal probability at any given point of the genetic evolution.

Zero-penalty moves. In order to increase the diversity of the children that a given parents can help create, we introduce zero-penalty moves for one of the parents that enter in any crossover operation. These moves do not alter the energy of that parent structure (hence the term *zero-penalty*), nor do they change its physical structure: a zero-penalty move simply changes the relative positioning of one of the parents relative to the other before a crossover between the two is performed. The specific moves that we use as zero-penalty "mutations" are (i) rotations around the axis by an arbitrarily chosen angle, (ii) axial displacements of the structure by random z values through the periodic boundary conditions, and (iii) 180° rotations around an arbitrary axis that is *perpendicular* to the z-axis. To clearly reveal the effect of zero-penalty moves in creating diversity in the genetic pool, let us consider the extreme case of a "generation zero" that is made up of only *identical* structures. With such a starting point for the GA run, any crossover is bound to create a child structure identical to the already existing members in the pool. However, if a parent is to be, for example, rotated about its axis before entering in a crossover operation, then that parent appears as distinct from the rest of the genetic pool and the new structure created is different from any other structure in the pool. Therefore, zero-penalty moves create the diversity necessary to a successful GA optimization even in cases where such diversity is completely lacking. When sufficient diversity is already present, a zero-penalty move does not hurt the performance of the algorithm because its computational cost is insignificant (the energy of the structure is never computed after such move).

With these descriptions of the ingredients of the algorithm, the procedural steps in our typical GA evolution are as follows:

a) Two parents are randomly picked from the genetic pool.
b) One of the parents is subjected to a zero-penalty move chosen with equal probability from the set (i)–(iii) described above.
c) After the zero-penalty move on the first parent, a crossover operation is performed with the (unchanged) second parent.
d) The resulting child is subject to a conjugate gradient relaxation into the nearest local minimum of the potential energy, and its "fitness" (potential energy per particle) is computed.
e) The child is selected for inclusion in the pool or rejected.
f) If the child does end up in the pool, then the members of the pool are sorted from the lowest to the highest energy per particle. One can easily implement and test variations of this procedure, which, for example, may use zero-penalty moves on both parents before they enter into a crossover. The cycle (a)–(f) is repeated for a prescribed number of crossovers or until at least the lowest-energy member of the pool has converged to a number of atoms and a value of the fitness function (energy per atom). The results from GA runs based on single crossover types (individual moves) and those based on all crossovers are presented in Section 5.3.2.3.

5.3.2.3 Results for Prototype Nanotubes

We have tested the genetic algorithm for two systems, carbon and LJ systems subjected to radial confinement conditions as described in the previous section. The purpose is to find out if nanotubes can evolve via genetic operations from structures that have one or few atoms in the periodic unit cell. To this end, we start the GA runs for the CNT systems from genetic pools in which each member has a single carbon atom in the periodic cell. We follow the evolution of GA runs in which only one type of crossover is employed from the set $\{S_1, S_2, S_3, P_1, P_2\}$ shown in Figure 5.30, as well as one GA run where *all* these operations are attempted *with equal probability*.

The evolution of the lowest energy in the pool for each of the six GA runs (i.e., five runs based on a single type of crossover and one run with all equiprobable crossovers) is plotted in Figure 5.31a, and the average energy across the pool is

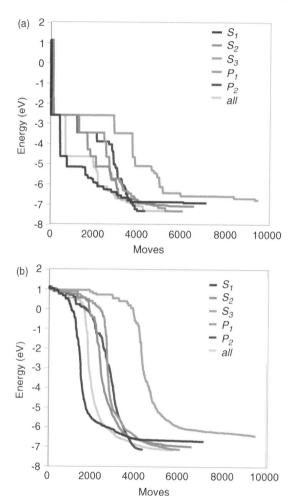

Figure 5.31 Evolution of (a) the lowest energy in the pool and (b) the average energy across the pool for CNT systems in separate runs performed with only one type of crossover, as well as in a run performed with all crossovers attempted with equal probability. The horizontal axis of each plot shows the number of crossovers (moves) attempted. Note that the GA runs based solely on sine operations could *not* find the global minimum structure within 10 000 crossovers. (From Ref. [13], with permission from Taylor and Francis.)

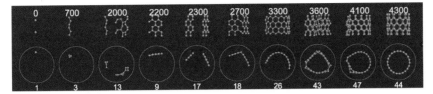

Figure 5.32 Evolution of lowest-energy carbon structure during a GA run performed with all crossover types. The system starts with only one atom for each member of the genetic pool and evolves toward a defect-free CNT as the lowest-energy pool member. The outer confining wall is shown by the white dashed line and the "time" (i.e., the index of the crossover operation) is indicated by the number shown atop each frame. The number of atoms is shown at the bottom of the frames; for clarity, two periodic lengths in the z-direction are displayed. (From Ref. [13], with permission from Taylor and Francis.)

shown in Figure 5.31b for each run. We observe that only the runs based on planar cuts (P_1 and P_2) are able to find the optimal defect-free CNT structure in less than 10^4 operations. The run based on all operations finds the correct structure in less than 5000 moves (crossovers). Since the individual sine crossovers are slow in finding the global optimum structure, the success of the all-move GA run is most likely due to the planar crossovers. The GA runs that are based on sine crossovers evolve their best structures to defective CNTs. A question remains as to whether performing longer runs with single crossovers of the sine type would eventually yield the correct defect-free tubular structures.

The structure of the best member in the genetic pool for the all-crossover GA runs on CNT systems is shown at selected time points in the evolution in Figure 5.32. As specified, the run started with a single carbon atom for every member of the pool, but we show two unit cells for each frame to help visualization through periodic boundary conditions in the z-direction. The best structure grows from one atom to a string of atoms that spans the length of the periodic cell (frame labeled 700 in Figure 5.32). A rather large number of crossovers have to be attempted in order for the string to grow wider (frames 2000 and 2200), that is, into a strip of sp^2-hybridized carbon atoms (graphene) at frame 2200. Planar crossovers performed with parent structures that consist of graphene strips will likely lead to two flat strips, as shown in frame 2300. The two strips subsequently coalesce at an angle (frame 2700 in Figure 5.32), acquire more atoms, and start curving onto a cylindrical surface due to the confining potential walls (frame 3300). Planar mating of parent structures such as that shown in frame 3300 results in closing the circumference, as shown in frame 3600; this closing occurs with defects along the crossover planes, but such defects are systematically weeded out later on (see frames 4100 and 4300).

We have also tested the GA for growing LJ nanotubes under radial confinement. To save time at the initial phase of particle accumulation, we have started the LJ runs with 10 atoms in each member of the GA pool located at random positions within the periodic unit cell. The evolution of the best energy in the pool for LJ systems (again, with individual crossovers and with all crossovers) is plotted in Figure 5.33a, and the average energy across the genetic pool is shown in Figure 5.33b. We note that for the LJ system, each of the GA operations is able to retrieve the correct

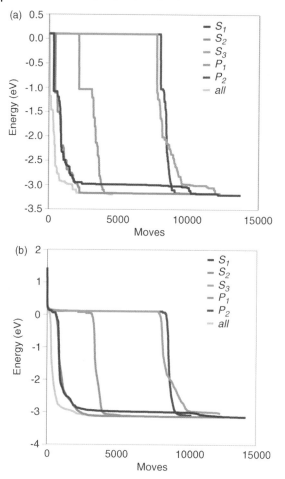

Figure 5.33 Evolution of (a) the lowest energy and (b) the average energy in the pool for the LJ system in separate runs performed with only one type of crossover, as well as in a run performed with all crossovers attempted with equal probability. (From Ref. [13], with permission from Taylor and Francis.)

optimum of the LJ nanotube compatible to the boundary and confinement conditions, although the fastest (S_2) and the slowest (S_1) runs are approximately 10 000 crossover operations apart.

The lowest-energy structure in the LJ GA pool for the all-crossover runs is shown in Figure 5.34 at selected points during its evolution. Interestingly, the evolution of the best LJ structure is rather smooth, that is, there are mainly accumulation events in which one or more cylindrical sectors of atoms grow larger. The coalescence of cylindrical pieces occurs with fewer defects than in the case of CNTs, and the completion of the circumference occurs about twice as fast as in the all-crossover CNT run. However, the elimination of point defects in the LJ nanotube structure is extremely slow. This comparison between the LJ and CNT systems points out to a

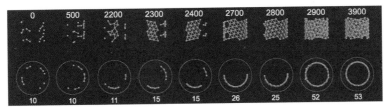

Figure 5.34 Evolution of lowest-energy LJ nanotube during a GA run with all crossover types. The system starts with 10 atoms for each member of the genetic pool and evolves toward a defect-free, helical LJ nanotube as the lowest-energy pool member. The outer confining wall is shown by the white dashed line and the "time" (i.e., index of the crossover operation) is indicated by the number shown atop each frame. The number of atoms is shown at the bottom of the frames; for clarity, two periodic lengths in the z-direction are displayed. (From Ref. [13], with permission from Taylor and Francis.)

strong influence of the interaction potential on the performance of the algorithm, which will be discussed in Section 5.3.2.4.

We present the evolution of the number of atoms in the lowest-energy member of the genetic pool in Figure 5.35. By comparing Figures 5.31 and 5.33 with Figure 5.35, we note the convergence of the energy per particle and the number of atoms is intrinsically related. For the CNT systems, the GA runs that do not find the lowest energy per particle also do not retrieve the correct number of atoms (i.e., the runs based on sine crossovers). Conversely, the runs that do find the lowest-energy structure also find the correct number of atoms (i.e., the runs based on planar cuts, and those using all operations). For the LJ systems, all runs are successful in finding the lowest-energy configuration *and* the optimal number of atoms. This comparison (i.e., Figures 5.31, 5.33, and 5.35) justifies, albeit *a posteriori*, our choice to use a variable-number genetic algorithm to solve the problem of finding 1D structures under cylindrical confinement.

5.3.2.4 Discussion

The results described in Section 5.3.2.3 for two prototype systems and for GA runs with different crossover operations allow us to draw useful conclusions on several issues:

The evolution of the number of atoms. The increase in number of atoms is facilitated by the algorithmic requirement that a newly created child structure C with an energy value equal to that of another member M of the population (parent or not) is nonetheless included in the pool if it has a different number of atoms from the member M. The result is that all the members of the pool are different in either energy or number of atoms, but not necessarily both. This requirement is crucial in the initial stage, as it is a very efficient way to allow the system to grow from one single atom. Most of the operations at this initial stage will combine 1-atom parents with zero or negligible energy into child structures with more atoms but still with negligible energy: operations with such result will continue until interatomic bonds are created and the fitness function criterion takes over in the selection process.

The effectiveness of a given crossover. In any genetic algorithm, the number of child acceptance events decreases as the algorithm progresses. In addition, we note

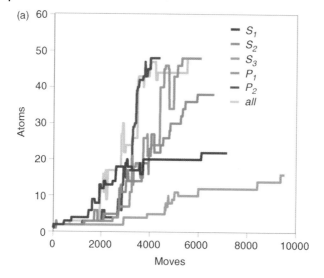
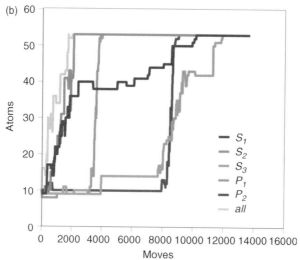

Figure 5.35 The evolution of the number of atoms corresponding to the lowest-energy member of the pool in the case of (a) CNT and (b) LJ systems, for GA runs performed with only one type of crossover and for a run performed with all crossovers. (From Ref. [13], with permission from Taylor and Francis.)

here that the efficiency of crossover operations depends on the particular physical system, as illustrated by the fact that the S_i operations ($i = 1, 2, 3$) are not effective for the CNT systems but do work during the LJ nanotube GA runs. For a given system, the *relative* efficiency of the different kinds of crossovers changes during the evolution. To illustrate this point, we look at the acceptance ratio of the moves during the GA runs with *individual* crossovers. We define the acceptance ratio at time n_{op} of the evolution (i.e., after a

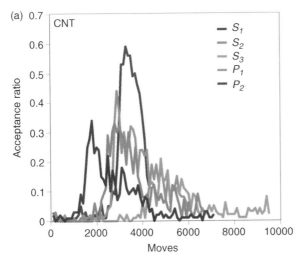

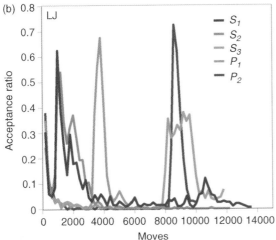

Figure 5.36 Acceptance ratios for single-crossover GA runs performed on the (a) CNT system and (b) the LJ system. (From Ref. [13], with permission from Taylor and Francis.)

total of n_{op} attempted moves) as the percentage of accepted children resulted from the 200 attempted moves that follow n_{op}. This value of 200 is chosen by numerical experimentation so that it is sufficiently small compared to the total duration of a GA run, but large enough to attenuate the statistical noise. The acceptance ratio is plotted in Figure 5.36a and b for the runs with single crossovers for the CNT and the LJ systems, respectively. Figure 5.36a for CNT systems shows that, initially, there is negligible acceptance rate for any crossover. Afterward, the S_1 operation is the most effective within the first 2000 operations of the evolution – then it decays. The acceptance ratios for the other single-crossover GA runs of the CNT system show similar peaks followed by decays, but, interestingly, the peaks are at different times and the subsequent decays occur with different rates. For example, the peaks of S_1, P_2, and S_3 are clearly distinct from one another in the CNT runs, whereas S_2 and P_1 are very similar (refer to Figure 5.36a). The fact that the acceptance ratio for individual crossovers peaks can occur at different times suggests the possibility *of using specific crossover types for different stages of the evolution* in order to accelerate the overall convergence of the algorithm. It is particularly important that the S_3 operation that mixes parents along their circumference and along their length reaches its acceptance ratio peak late and has longer tail: the intent of this move was to create changes on small areas of the tube (and thus, to serve as a replacement for mutations). Clearly, the S_3 move should be used mostly in the late stages of the evolution where the system is close to the correct structures except for localized defects that can be directly affected by S_3 operations. For the LJ system, Figure 5.36b shows very similar trends for all crossover operations in that negligible acceptance ratios are followed by significant peaks and then by gradual decays, with the main difference that the peaks that are better separated may belong to different crossover types than in the case of CNTs, and the time separation (number of operations) between them is very different than in the CNT case. The S_3 move still peaks up late, indicating again that mutations can be effective at the late stages of the evolution; however, for the LJ system, both S_3 and P_2 have peaks at the late stages, with the P_2 peak being of higher value.

Ideally, a very large number of runs should be performed in order to draw quantitative conclusions about the occurrence of the acceptance ratio peaks and the nature of their decay. We have not performed sufficient GA runs to generate reliable statistics of the acceptance ratios for the main reason that the actual time taken by any given run is rather long, on the order of couple of days on a single processor. Still, we have repeated every run for the CNT and LJ systems exactly three times and found that (a) the lowest-energy structure and its corresponding number of atoms did not change, (b) and acceptance ratios behaved as peaks followed by decays during each run. However, we noticed that the highest peaks in acceptance ratios occurred at widely different times during the evolution of different GA runs for the same move used, especially for the runs started from a single atom for each pool member. This observation makes the question of statistics a rather expensive one, for which reason such question will be deferred to future studies.

The equiprobable combination of crossovers. The all-crossover runs for LJ and CNT systems reach their corresponding optimum structures in less than 5000 operations. The operations used are the same for both systems and they provide *sufficient structural diversity* for the all-crossover GA to be successful. Each crossover type is attempted with equal probability. The acceptance ratio for a given crossover at a certain time in the all-move evolution is defined as the number of accepted children created by that crossover divided by the total number of accepted children created by all operations during an interval of 200 crossovers following that time. When comparing the performance of types of crossovers within the all-move GA runs, we note that the acceptance ratios of the different crossovers peak at similar times, and decay over similar periods (refer to Figure 5.37). No one particular move could be

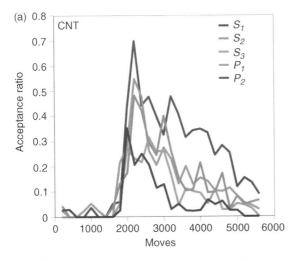

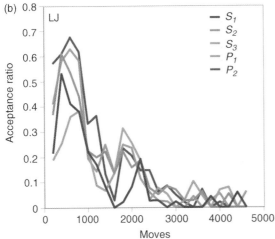

Figure 5.37 Acceptance ratios for each crossover during an all-crossover GA run for (a) the CNT system and (b) the LJ nanotubes. (From Ref. [13], with permission from Taylor and Francis.)

clearly assigned as responsible for the steady progress of the algorithm when the others have low acceptance ratios. So from the point of view of convergence speed, the equiprobable combination of moves is not ideal. From that standpoint alone, the equiprobable approach is actually somewhat inefficient, because both the moves that are likely to succeed at a given stage of the evolution and those moves that are not are given equal chance. However, speed is a price paid for robustness, because the many crossovers attempted with equal probability make the algorithm work for very different physical systems with no change whatsoever. Even though the algorithm retrieves the optimal configurations for the CNT and LJ systems using the very same set of crossovers and attempting similar numbers of moves, it does so going through very different structural stages (refer to Figures 5.32 and 5.34). For example, the LJ nanotube is completed very fast, and it is only the stage of eliminating the defects (frames 2900 through 3900) that brings the run closer in length to the CNT (which succeed after 4400 moves). That the interaction model does affect the performance of a run can be seen more clearly and directly from the single-crossover GA runs: for example, S_1 succeeds for LJ nanotubes, but does not succeed for CNTs. The influence of the interaction models on the performance of the GA is clearly diminished by allowing many different types of crossovers to be attempted during a single run.

5.3.2.5 Concluding Remarks

In conclusion, we have presented a genetic algorithm for growing tubular nanostructure subjected to periodic boundary conditions and radial confinement. The main features of this algorithm are (a) the selection process that allows the number of atoms to vary over a wide range, (b) the use of several different types of crossovers in combinations in which they have the same attempt rate, and (c) the lack of explicit mutations. With this algorithm, nanotubes can grow into their global minimum (at given constraints) starting from one or few atoms. Both the optimal number of atoms and their configuration are found within the same GA search. The presence of several crossovers makes the algorithm robust and readily applicable for growing different 1D materials systems. We have analyzed the acceptance of crossovers in single- and in all-move runs and showed that the equiprobable combination of moves makes the global minimum attainable for CNT and LJ nanotubes but does not improve the convergence speed *per se*. From the comparative study of the CNT and the LJ systems, it is clear that the using only one type of move will not always lead to the global minimum. The performance of each crossover is different at various stages of the genetic evolution, and investigations focused on improving the convergence speed of all-crossover runs are underway.

Another interesting future step would be to develop genetic algorithms for growth of nanostructures with an optimized desired property of technological interest (e.g., resistivity, thermal conductivity, and chemical activity) [123,124] *in addition* to using a physical fitness function (e.g., energy) for selection of the members of the genetic pool. Such technologically oriented problems that are still linked to an actual synthesis process will require the simultaneous optimization of several objective functions, and particular complexity arises when these objective functions have a competitive nature [125].

References

1 Deaven, D.M. and Ho, K.M. (1995) *Phys. Rev. Lett.*, **75**, 288.
2 Chuang, F.C., Ciobanu, C.V., Shenoy, V.B., Wang, C.Z., and Ho, K.M. (2004) *Surf. Sci. Lett.*, **573**, L375.
3 Abraham, N.L. and Probert, M.I.J. (2006) *Phys. Rev. B*, **65**, 224104.
4 Oganov, A.R. and Glass, C.W. (2006) *J. Chem. Phys.*, **124**, 244704.
5 Trimarchi, G. and Zunger, A. (2007) *Phys. Rev. B*, **75**, 104113.
6 Ciobanu, C.V. and Predescu, C. (2004) *Phys. Rev. B*, **70**, 085321.

7 Chuang, F.C., Ciobanu, C.V., Wang, C.Z., and Ho, K.M. (2005) *J. Appl. Phys.*, **98**, 073507.
8 Chuang, F.C., Ciobanu, C.V., Predescu, C., Wang, C.Z., and Ho, K.M. (2005) *Surf. Sci.*, **578**, 183.
9 Zang, J., Wang, C.Z., and Ho, K.M. (2009) *Phys. Rev. B*, **80**, 174102.
10 Briggs, R.M. and Ciobanu, C.V. (2007) *Phys. Rev. B*, **75**, 195415.
11 Chan, T.L., Ciobanu, C.V., Chuang, F.C., Lu, N., Wang, C.Z., and Ho, K.M. (2006) *Nano Lett.*, **6**, 277.
12 Lu, N., Ciobanu, C.V., Chan, T.L., Chuang, F.C., Wang, C.Z., and Ho, K.M. (2007) *J. Phys. Chem. C*, **111**, 7933.
13 Davies, T.E.B., Mehta, D.P., Rodriguez-Lopez, J.L., Gilmer, G.H., and Ciobanu, C.V. (2009) *Mater. Manuf. Process.*, **24**, 265.
14 Novoselov, K.S., Jiang, D., Schedin, F., Booth, T.J., Khotkevich, V.V., Morozov, S.V., and Geim, A.K. (2005) *Proc. Natl. Acad. Sci. USA*, **102**, 10451.
15 Ranke, W. and Xing, Y.R. (1997) *Surf. Sci.*, **381**, 1.
16 Suzuki, T., Minoda, H., Tanishiro, Y., and Yagi, K. (1996) *Surf. Sci.*, **358**, 522; Suzuki, T., Minoda, H., Tanishiro, Y., and Yagi, K. (1996) *Surf. Sci.*, **348**, 335.
17 Liu, J., Takeguchi, M., Yasuda, H., and Furuya, K. (2002) *J. Cryst. Growth*, **237–239**, 188.
18 Joeng, S., Jeong, H., Cho, S., and Seo, J.M. (2004) *Surf. Sci.*, **557**, 183.
19 Baski, A.A., Erwin, S.C., and Whitman, L.J. (1995) *Science*, **269**, 1556.
20 Lee, G.D. and Yoon, E. (2003) *Phys. Rev. B*, **68**, 113304.
21 Woodruff, D.P. (2002) *Surf. Sci.*, **500**, 147.
22 Kresse, G., Bergermayer, W., Podloucky, R., Lundgren, E., Koller, R., Schmid, M., and Varga, P. (2003) *Appl. Phys. A*, **76**, 701.
23 Radecke, M. and Carter, E.A. (1995) *Phys. Rev. B*, **51**, 4388.
24 Fujikawa, Y., Akiyama, K., Nagao, T., Sakurai, T., Lagally, M.G., Hashimoto, T., Morikawa, Y., and Terakura, K. (2002) *Phys. Rev. Lett.*, **88**, 176101.
25 Raiteri, P., Migas, D.B., Miglio, L., Rastelli, A., and von Känel, H. (2002) *Phys. Rev. Lett.*, **88**, 256103.
26 Shenoy, V.B., Ciobanu, C.V., and Freund, L.B. (2002) *Appl. Phys. Lett.*, **81**, 364.
27 Ciobanu, C.V., Shenoy, V.B., Wang, C.Z., and Ho, K.M. (2003) *Surf. Sci.*, **544**, L715.
28 Migas, D.B., Cereda, S., Montalenti, F., and Miglio, L. (2004) *Surf. Sci.*, **556**, 121.
29 Mo, Y.W., Savage, D.E., Swartzentruber, B.S., and Lagally, M.G. (1990) *Phys. Rev. Lett.*, **65**, 1020.
30 Lenosky, T.J., Sadigh, B., Alonso, E., Bulatov, V.V., Diaz de la Rubia, T., Kim, J., Voter, A.F., and Kress, J.D. (2000) *Model. Simul. Mater. Sci. Eng.*, **8**, 825.
31 Zhao, R.G., Gai, Z., Li, W., Jiang, J., Fujikawa, Y., Sakurai, T., and Yang, W.S. (2002) *Surf. Sci.*, **517**, 98.
32 Gai, Z., Yang, W.S., Zhao, R.G., and Sakurai, T. (1999) *Phys. Rev. B*, **59**, 13003.
33 Seehofer, L., Bunk, O., Falkenberg, G., Lottermoser, L., Feidenhans'l, R., Landemark, E., Nielsen, M., and Johnson, R.L. (1997) *Surf. Sci.*, **381**, L614.
34 Gai, Z., Zhao, R.G., Ji, H., Li, X., and Yang, W.S. (1997) *Phys. Rev. B*, **56**, 12308.
35 Zhao, R.G., Gai, Z., Lo, W., Jiang, J., Fujikawa, Y., Sakurai, T., and Yang, W.S. (2002) *Surf. Sci.*, **517**, 98.
36 Wu, Y.Q., Li, F.H., Cui, J., Lin, J.H., Wu, R., Qin, J., Zhu, C.Y., Fan, Y.L., Yang, X.J., and Jiang, Z.M. (2005) *Appl. Phys. Lett.*, **87**, 223116.
37 Chuang, F.C., Ciobanu, C.V., Shenoy, V.B., Wang, C.Z., and Ho, K.M., (2004) *Surf. Sci.*, **573**, L375; Zandvliet, H.J.W. (2005) *Surf. Sci.*, **577**, 93.
38 Chuang, F.C., Ciobanu, C.V., Predescu, C., Wang, C.Z., and Ho, K.M. (2005) *Surf. Sci*, **578**, 183.
39 Chuang, F.C., Ciobanu, C.V., Wang, C.Z., and Ho, K.M. (2005) *J. Appl. Phys.*, **98**, 073507.
40 Fujikawa, Y., Akiyama, K., Nagao, T., Sakurai, T., Lagally, M.G., Hashimoto, T., Morikawa, Y., and Terakura, K., (2002) *Phys. Rev. Lett.*, **88**, 176101; Raiteri, P., Migas, D.B., Miglio, L., Rastelli, A., and von Känel, H., (2002) *Phys. Rev. Lett.*, **88**, 256103; Shenoy, V.B., Ciobanu, C.V., and Freund, L.B. (2002) *Appl. Phys. Lett.*, **81**, 364.
41 Khor, K.E. and Das Sarma, S. (1997) *J. Vac. Sci. Technol. B*, **15**, 1051.
42 Cereda, S., Montalenti, F., and Miglio, L. (2005) *Surf. Sci.*, **591**, 23.
43 Erwin, S.C., Baski, A.A., and Whitman, L.J. (1996) *Phys. Rev. Lett.*, **77**, 687.

44 Tersoff, J. (1989) *Phys. Rev. B*, **39**, 5566.
45 Ciobanu, C.V., Shenoy, V.B., Wang, C.Z., and Ho, K.M. (2003) *Surf. Sci.*, **544**, L715.
46 Seehofer, L., Falkenberg, G., and Johnson, R.L., (1996) *Phys. Rev. B*, **54**, R11062; Chuang, F.C. (2007) *Phys. Rev. B*, **75**, 115408.
47 So far it appears that only high-coverage, high-deposition rate experiments have been reported in the Ge/Si(103) heteroepitaxial system: Gai, Z., Yang, W.S., Sakurai, T., and Zhao, R.G. (1999) *Phys. Rev. B*, **59**, 13009.
48 Hu, X.M., Wang, E.G., and Xing, Y.R. (1996) *Appl. Surf. Sci.*, **103**, 217.
49 Baski, A.A., Erwin, S.C., and Whitman, L.J. (1995) *Science*, **269**, 1556.
50 Liu, J., Takeguchi, M., Yasuda, H., and Furuya, K. (2002) *J. Cryst. Growth*, **237**, 188.
51 Jeong, S., Jeong, H., Cho, S., and Seo, J.M. (2004) *Surf. Sci.*, **557**, 183.
52 Baski, A.A., Erwin, S.C., and Whitman, L.J. (1997) *Surf. Sci.*, **392**, 69.
53 Ranke, W. and Xing, Y.R. (1985) *Phys. Rev. B*, **31**, 2246.
54 Gardeniers, J.G.E., Maas, W.E.J.R., van Meerten, R.Z.C., and Giling, L.J. (1989) *J. Cryst. Growth*, **96**, 821.
55 Baski, A.A. and Whitman, L.J. (1995) *Phys. Rev. Lett.*, **74**, 956.
56 Baski, A.A. and Whitman, L.J. (1995) *J. Vac. Sci. Technol. A*, **13**, 1469.
57 Baski, A.A. and Whitman, L.J. (1996) *J. Vac. Sci. Technol. B*, **14**, 992.
58 Song, S., Yoon, M., and Mochrie, S.G.J. (1995) *Surf. Sci.*, **334**, 153.
59 Takeguchi, M., Wu, Y., and Furuya, K. (2000) *Surf. Interface Anal.*, **30**, 288.
60 Ranke, W. and Xing, Y.R. (1997) *Surf. Rev. Lett.*, **4**, 15.
61 Quantum Espresso, http://www.quantum-espresso.org/, last accessed January 2nd, 2013.
62 Erwin, S.C., Baski, A.A., and Whitman, L.J. (1996) *Phys. Rev. Lett.*, **77**, 687.
63 Dabrowski, J., Müssig, H.J., and Wolff, G. (1994) *Phys. Rev. Lett.*, **73**, 1660.
64 Ciobanu, C.V., Tambe, D.T., Shenoy, V.B., Wang, C.Z., and Ho, K.M. (2003) *Phys. Rev. B*, **68**, 201302.
65 Riikonen, S. and Sanchez-Portal, D. (2005) *Nanotechnology*, **16**, S218.
66 Crain, J.N., McChesney, J.L., Zheng, F., Gallagher, M.C., Snijders, P.C., Bissen, M., Gundelach, C., Erwin, S.C., and Himpsel, F.J. (2004) *Phys. Rev. B*, **69**, 125401.
67 Owen, J.H.G., Miki, K., and Bowler, D.R. (2006) *J. Mater. Sci.*, **41**, 4568.
68 Burton, W.K., Cabrera, N., and Frank, F. (1951) *Philos. Trans. R. Soc. Lond. A*, **243**, 299.
69 Laracuente, A., Erwin, S.C., and Whitman, L.J. (1999) *Appl. Phys. Lett.*, **74**, 1397.
70 Poon, T.W., Yip, S., Ho, P.S., and Abraham, F.F. (1992) *Phys. Rev. B*, **45**, 3521.
71 Zandvliet, H.J.W. (2000) *Rev. Mod. Phys.*, **72**, 593.
72 Chadi, D.J. (1987) *Phys. Rev. Lett.*, **59**, 1691.
73 Hornstra, R. (1959) *Physica*, **25**, 409.
74 Grovenor, R.M. (1985) *J. Phys. C*, **18**, 4079.
75 Arias, T.A. and Joannopoulos, J.D. (1994) *Phys. Rev. B*, **49**, 4525.
76 Maiti, A., Chisholm, M.F., Pennycook, S.J., and Pantelides, S.T. (1996) *Phys. Rev. Lett.*, **77**, 1306.
77 Kohyama, M. (1987) *Phys. Status Solidi B*, **141**, 71.
78 Kohyama, M., Yamamoto, R., Ebata, Y., and Kinoshita, M. (1988) *J. Phys. C*, **21**, 3205.
79 Kohyama, M., Yamamoto, R., and Doyama, M. (1986) *Phys. Status Solidi B*, **137**, 11.
80 Paxton, T. and Sutton, A.P. (1988) *J. Phys. C*, **21**, L481.
81 Levi, A., Smith, D.A., and Weizel, J.T. (1991) *J. Appl. Phys.*, **69**, 2048.
82 Sutton, A.P. Vitek, V. (1983) *Philos. Trans. R. Soc. Lond. A*, **309**, 1.
83 Morris, J.R., Lu, Z.Y., Ring, D.M., Xiang, J.B., Ho, K.M., Wang, C.Z., and Fu, C.L. (1998) *Phys. Rev. B*, **58**, 11241.
84 Shenderova, O.A., Brenner, D.W., and Yang, L.H. (1999) *Phys. Rev. B*, **60**, 7043.
85 Rouviere, J.L. and Bourret, A. (1989) *Polycrystalline Semiconductors, Springer Proceedings in Physics*, vol. 35 (eds H.J. Möller, H.P. Strunk, and J.H. Werner), Springer, Berlin, p. 19.
86 Rouviere, J.L. and Bourret, A. (1990) *J. Phys.*, **51**, Cl–C329.

87 Chisholm, M.F. and Pennycook, S.J. (1997) *MRS Bull.*, **22**, 53.
88 Tang, M.S., Wang, C.Z., Chan, C.T., and Ho, K.M. (1996) *Phys. Rev. B*, **53**, 979.
89 Appell, D. (2002) *Nature*, **419**, 553.
90 Cui, Y. and Lieber, C.M. (2001) *Science*, **291**, 851.
91 Morales, A.M. and Lieber, C.M. (1998) *Science*, **269**, 208.
92 Zhang, R-.Q., Lifshitz, Y., and Lee, S-.T. (2003) *Adv. Mater.*, **15**, 635.
93 Holmes, J.D., Johnston, K.P., Doty, R.C., and Korgel, B.A. (2000) *Science*, **287**, 1471.
94 Ma, D.D.D., Lee, C.S., Au, F.C.K., Tong, S.Y., and Lee, S.T. (2003) *Science*, **299**, 1874.
95 Wu, Y., Cui, Y., Huynh, L., Barrelet, C.J., Bell, D.C., and Lieber, C.M. (2004) *Nano Lett.*, **4**, 433.
96 Schmidt, V., Senz, S., and Gösele, U. (2005) *Nano Lett.*, **5**, 931.
97 Menon, M. and Richter, E. (1999) *Phys. Rev. Lett.*, **83**, 792.
98 Zhao, Y. and Yakobson, B.I. (2003) *Phys. Rev. Lett.*, **91**, 035501.
99 Bai, J., Zeng, X.C., Tanaka, H., and Zeng, J.Y. (2004) *Proc. Natl. Acad. Sci. USA*, **101**, 2664.
100 Rurali, R. and Lorente, N. (2005) *Phys. Rev. Lett.*, **94**, 026805.
101 Kagimura, R., Nunes, R.W., and Chacham, H. (2005) *Phys. Rev. Lett.*, **95**, 115502.
102 Cui, Y., Gudiksen, L.J., Wang, M.S., and Lieber, C.M. (2001) *Appl. Phys. Lett.*, **78**, 2214.
103 Wang, N., Tang, Y.H., Zhang, Y.F., Lee, C.S., Bello, I., and Lee, S.T. (1999) *Chem. Phys. Lett.*, **299**, 237.
104 Hanrath, T. and Korgel, B.A. (2005) *Small*, **1**, 717.
105 Pimpinelli, A. and Villain, J. (1998) Chapter 3, in *Physics of Crystal Growth*, Cambridge University Press.
106 Ho, K.M., Shvartsburg, A.A., Pan, B.C., Lu, Z.Y., Wang, C.Z., Wacker, J.G., Fye, J.L., and Jarrold, M.F. (1998) *Nature*, **392**, 582.
107 Hansen, U. and Vogl, P. (1998) *Phys. Rev. B*, **57**, 13295.
108 VIENNA *ab initio* simulation package, Technische Universität Wien, 1999; Kresse, G. and Hafner, J., (1993) *Phys. Rev. B*, **47**, R558; Kresse, G. and Furthmuller, J. (1996) *Phys. Rev. B*, **54**, 11169.
109 Perdew, J.P. (1991) in *Electronic Structure of Solids '91* (eds P. Ziesche and H. Eschrig), Akademie-Verlag, Berlin.
110 Zhao, X., Wei, C.M., Yang, L., and Chou, M.Y. (2004) *Phys. Rev. Lett.*, **92**, 236805.
111 Iijima, S. (1991) *Nature*, **354**, 56.
112 Li, H., Pederiva, F., Wang, G.H., and Wang, B.L. (2004) *J. Appl. Phys.*, **96**, 2214.
113 Wang, B.L., Wang, G.H., and Zhao, J.J. (2002) *Phys. Rev. B*, **65**, 235406.
114 Chan, T.L., Ciobanu, C.V., Chuang, F.C., Lu, N., Wang, C.Z., and Ho, K.M. (2006) *Nano Lett.*, **6**, 277.
115 Lu, N., Ciobanu, C.V., Chan, T.L., Chuang, F.C., Wang, C.Z., and Ho, K.M. (2007) *J. Phys. Chem. C*, **111**, 7933.
116 Sastry, K., Goldberg, D.E., and Johnson, D.D. (2007) *Mater. Manuf. Process.*, **22**, 570.
117 Chakraborti, N., Jayakanth, R., Das, S., Calisir, E.A., and Erkoc, S. (2007) *J. Phase Equilib. Diff.*, **28**, 140.
118 Call, S.T., Zubarev, D.Y., and Boldyrev, A.I. (2007) *J. Comput. Chem.*, **28**, 1177–1186.
119 Allen, M.P. and Tildesley, D.J. (1987) *Computer Simulation of Liquids*, Clarendon Press, Oxford.
120 The relative positions of the two repulsive walls are chosen such that the interaction potential exerted by the wall does not significantly affect the energy of the nanotube created in between the two walls. The repulsion potential of each wall has a form similar to that given in Sabo, D., Predescu, C., Doll, J.D., and Freeman, D.L. (2004) *J. Chem. Phys.*, **121**, 856.
121 Tersoff, J. (1989) *Phys. Rev. B*, **39**, 5566.
122 Lennard-Jones, J.E. and Ingham, A.E. (1925) *Proc. R. Soc. Lond. A*, **107**, 636.
123 Fumiyoshi, Y., Shin-Ichi, F., Suchada, W., and Mitsuru, H. (2006) *J. Drug Target.*, **14**, 496.
124 Chen, Y., Li, D., Lukes, J.R., and Majumdar, A. (2005) *J. Heat Transfer*, **127**, 1129.
125 Fonseca, C.M. and Fleming, P.J. (2000) Multiobjective optimization, in *Evolutionary Computation 2* (eds T. Back, D.B. Fogel, and Z. Michalewicz), Institute of Physics Publishing, Bristol, 270 pp.

6
Other Methodologies for Investigating Atomic Structure

The importance of structure, especially in the context of ground state of clusters and crystal structure prediction, has sparked long and intense investigations in diverse algorithms and applications. These investigations resulted in a wide range of techniques, of which so far in this book we dealt only with genetic algorithms. We enumerate and discuss in this chapter a few other techniques; given that the genetic/evolutionary techniques are the explicit scope of this book, we describe in some detail only few optimization techniques in the subsequent sections (Sections 6.1–6.4). Thus, while we are clearly not doing justice to the depth and sophistication of these other global optimization techniques, we believe it is important to give the reader (most likely the eager graduate student) an idea of the breadth of the field of structure optimization by showing here some of the most powerful ones. For a wider and more detailed account of most of the current methods of crystal structure prediction, we refer the reader to the recent book edited by Oganov [1].

Recalling the discussion of Chapter 3, structure determination techniques have to include at some point local relaxation for the reason that this decreases the effective dimensionality of the search space by dealing with the reduced potential energy surface (PES). Exception is simulated annealing (SA) [2], or in its more robust or ergodic form, parallel tempering Monte Carlo (PTMC) (Section 6.1) simulated annealing, which deals with the full PES and only at the end relaxes the structures so that they can be classified in a way that does not depend on temperature.

For small systems with up to perhaps 10 atoms, even *random sampling* can furnish correct global minima when coupled with local optimization [3,4]. This technique, although straightforward and general, is the one that suffers the most from the curse of dimensionality because it lacks "directive" toward the global optimum and one is left to hope that there were sufficiently many structures visited and relaxed such that the global optimum was one of them. However, as argued in Chapter 3, this cannot be a feasible way to address larger systems.

Simulated annealing is the most simple way to simulate the actual thermodynamics of the system. If ground state is desired, then one can in principle equilibrate the system at high temperature and then cool slowly in order to have the system crystallize (hopefully) in the ground state. While the current wisdom in simulated annealing seems to be that moderate or large systems get trapped in local minima and can never be cooled slowly enough that the final state ends up being the ground state, we should point out that there are ways to ensure that the system becomes ergodic or very nearly so using reasonable computational time and resources. One such way is described at length in Section 6.1. The long-standing appeal of SA and, necessarily, in the near future of PTMC or other SA-related techniques is that

a) they are direct ways of simulating the actual thermodynamics of the systems, that is, thermodynamics based on the real PES and not on the reduced PES (which has, by construction, smaller energy barriers between minima) over a relevant range of temperatures;
b) they are easy and robust to implement.

As we can clearly see, PTMC (and any SA technique) is not designed for global minimum search, but can serve very well for this purpose provided that the problem of getting trapped in local minima is properly addressed. There are certain numbers of parameters that can be adjusted in SA [5] in order to avoid this trapping problem, the most conceptually important of them probably being some aptly designed form of temperature scheduling: in such schemes, one would attempt periodic exchanges between Monte Carlo walkers at high temperature and lower ones, in order to kick the ones at lower temperatures out of the local minima in which they could otherwise get easily trapped. Precise details about one way to achieve this are given in Section 6.1, with results described for a prototype crystalline surface, Si(114), in Section 6.5.1.

Another class of very powerful methods for structure prediction consists of methods based on the reduced PES, from which we mention the basin hopping Monte Carlo (BHMC) and the minima hopping (MH). These methods are discussed briefly in Sections 6.2 and 6.3. Since 1997, BHMC has proven an extremely versatile approach to identify global minima and low-energy structures especially for atomic clusters. In fact, the first feat obtained came with the original paper by Doye and Wales [6], in which global minima of Lennard-Jones clusters up to 110 were reported.

Recently, Goedecker combined the basin hopping method with a simulated annealing strategy to search for the global minimum in an optimization algorithm called minima hopping [7]. In this MH method, molecular dynamics (MD) simulations are performed between local structure optimizations. The MD process helps the system jump from one local minimum to another. Furthermore, Goedecker designed a feedback mechanism to control the MD temperature and acceptance probability of newly located minima. This algorithm has been applied to cluster structures and demonstrated good performance. This type of method has also been applied to periodic systems recently [8]. We review the MH approach to structure optimization in Section 6.3.

Although many of these search algorithms have been around for some time and have been applied successfully in many systems, there is still no clear understanding of what governs the efficiencies of these algorithms in various global structural searches. This is because the performance of a global optimization scheme strongly depends on the size of the problem and the nature of the landscape being explored and it is often difficult to translate performance on a given problem and size to another problem or another size of the same problem. For example, a given algorithm might work very well at small sizes but have a dramatic decrease in efficiency once the size of the system becomes larger. In particular, different implementation of structural representation and mating operation in GA could dramatically influence the algorithm behavior. If an algorithm is not well designed and lattice vectors are not chosen properly, we could be trapped in various local minima and the search efficiency will be reduced. In order to develop more efficient algorithms, benchmark calculations will be valuable.

In Sections 6.5.1 and 6.5.2, we examine the performance of our GA approach in comparison with PTMC for the surface reconstruction problem, in comparison with MH for crystal structure prediction. Several systems that have been studied successfully in previous works [9,10] will be revisited using a GA with a real-space representation. For comparison, we also applied the MH method on some of the systems. Since both methods relax candidates to local minima before selection, these methods, GA and MH, actually compare on equal footing and search over the same configuration space consisting of a collection of local basins but traverse the space in very different ways.

6.1
Parallel Tempering Monte Carlo Annealing

6.1.1
General Considerations

In choosing a methodology that can help predict the surface reconstructions, we have taken into account the following considerations. First, the number of atoms in the simulation slab is large because it includes several subsurface layers in addition to the surface ones. Moreover, the number of local minima of the potential energy surface is also large, as it scales roughly exponentially [11,12] with the number of atoms involved in the reconstruction; by itself, such scaling requires the use of fast stochastic search methods. Second, methods that are based on the modification of the potential energy surface (such as the basin hoping [13] algorithm), although very powerful in predicting global minima, have been avoided as our future studies are aimed at predicting not only the correct lowest-energy reconstructions, but also the thermodynamics of the surface. Finally, the calculation of interatomic forces is expensive, so the method should be based on Monte Carlo algorithms rather than molecular dynamics. We mention, however, that recent advances in molecular dynamics algorithms, especially the parallel replica [14] and temperature-accelerated dynamics [15] developed by Voter and coworkers, may constitute viable alternatives to Monte Carlo parallel tempering for the sampling of low-temperature systems.

These considerations, coupled with a desire for simplicity and robustness of implementation, prompted us to choose the parallel tempering Monte Carlo algorithm [16,17] for this study. Before describing in detail the computational procedure and its advantages, we pause to discuss the computational cell and the empirical potential used.

The simulation cell has a single-face slab geometry with periodic boundary conditions applied in the plane of the surface (denoted xy), and no periodicity in the direction (z) normal to the surface. The "hot" atoms from the top part of the slab (10–15 Å thick) are allowed to move, while the bottom layers of atoms are kept fixed to simulate the underlying bulk crystal. Though highly unlikely during the finite time of the simulation, the evaporation of atoms is prevented by using a wall of infinite energy that is parallel to the surface and situated 10 Å above it; an identical wall is placed at the level of the lowest fixed atoms to prevent the (theoretically possible) diffusion of the hot atoms through the bottom of the slab. The area of the simulation cell in the xy-plane and the number of atoms in the cell are kept fixed during each simulation; as we shall discuss in Section 6.5, these assumptions are not restrictive as long as we consider all the relevant values of the number of atoms per area. Under these conditions, the problem of finding the most stable reconstruction reduces to the global minimization of the total potential energy $V(\mathbf{x})$ of the atoms in the simulation cell (here $\mathbf{x}$ denotes the set of atomic positions). In order to sort through the numerous local minima of the potential $V(\mathbf{x})$, a stochastic search is necessary. The general strategy of such search (as illustrated, for example, by the simulated annealing technique [2,18]) is to sample the canonical Boltzmann distribution $exp[-V(\mathbf{x})/(k_B T)]$ for decreasing values of the temperature T and look for the low-energy configurations that are generated.

In terms of atomic interactions, we are constrained to use empirical potentials because the highly accurate *ab initio* or tight-binding methods are prohibitive. Since this work is aimed at finding the *lowest*-energy reconstructions for arbitrary surfaces, the choice of the empirical potential is crucial. There are two key issues that make an empirical potential suitable for finite-temperature simulations aimed at structure determination: (a) the potential has to reproduce (relaxed) ground-state structures that are known from experiments or from electronic structure calculations, and (b) the energetic barriers encountered by atoms moving at the surface have to be physically relevant. The first issue is usually reduced to checking that a known structure is retained upon relaxation [19]. A stronger and more meaningful test would be to rank order several reconstructions and check that the ordering is the same as that given by density functional methods. In our studies, we have found that there are serious deficiencies of the empirical potentials as far as surface structure is concerned. Two relevant examples are in order: (i) contrary to the common belief [19], the ground-state structure of the Si(001) surface as given by the Tersoff potential [20] is *not* the (2 × 1) reconstruction but rather a certain pattern of dimer rows that are in and out of phase (see also Ref. [21]), and (ii) the Stillinger–Weber potential [22] does not reproduce the Si(113) adatom interstitial reconstructions observed experimentally, that is, those structures are not even local minima.

To our knowledge, the recent work of Nurminen *et al.* [21] has been the first to point out that the widespread empirical potentials [20,22] are not very suitable for finite-temperature surface calculations, suggesting the issue of unphysical energy barriers at the surface. In practice, such barriers are not explicitly used in the fitting of empirical potentials that rely mostly on bulk-related properties for performing the fitting. There are, however, several remarkable efforts to include information about atomic clusters when fitting the parameters of empirical potentials (e.g., see Refs [23,24]). Therefore, potentials that include information about bulk and cluster structure are more likely to perform satisfactorily for surfaces. After numerical experimentation with several empirical models, we have chosen to use the highly optimized empirical potential (HOEP) recently developed by Lenosky et al. [25]. HOEP is fitted to a large database of *ab initio* calculations using the force-matching method and provides a good description of the energetics of all atomic coordinations up to $Z = 12$. Furthermore, we have tested that HOEP has superior transferability to the different kinds of bonding environments present on various high-index surfaces.

6.1.2
Advantages of the Parallel Tempering Algorithm as a Global Optimizer

The parallel tempering Monte Carlo method (also known as the replica-exchange Monte Carlo method) consists in running parallel canonical simulations of many statistically independent replicas of the system, each at a different temperature $T_1 < T_2 < \cdots < T_N$. The set of N temperatures $\{T_i, i = 1, 2, \ldots, N\}$ is called a *temperature schedule*, or *schedule* for short. The probability distributions of the individual replicas are sampled with the Metropolis algorithm [26], although any other ergodic strategy can be utilized. In particular, we note the possibility of using molecular dynamics to sample the configuration space of each individual replica (e.g., see Ref. [27]). Irrespective of what sampling strategy is being used for each replica, the key feature of the parallel tempering method is that swaps between replicas of neighboring temperatures T_i and T_j ($j = i \pm 1$) are proposed and allowed with the conditional probability [16,17] given by

$$\min\{1, e^{(1/T_j - 1/T_i)[V(\mathbf{x}_j) - V(\mathbf{x}_i)]/k_B}\}, \tag{6.1}$$

where $V(\mathbf{x}_i)$ represents the energy of the replica i and k_B is the Boltzmann constant. The conditional probability (6.1) ensures that the detailed balance condition is satisfied and that the equilibrium distributions are the Boltzmann ones for each temperature.

Because of the swapping mechanism, parallel tempering enjoys certain advantages (as a global optimizer) over the more popular simulated annealing algorithm [2,18]. In order for SA to be convergent (i.e., to reach the global optimum as the temperature is lowered), the cooling schedule must be of the form [28,29]

$$T_i = \frac{T_0}{\log(i + i_0)}, \quad i \geq 1, \tag{6.2}$$

where T_0 and i_0 are sufficiently large constants. Such a logarithmic schedule is too slow for practical applications, and faster schedules are routinely utilized. Common SA cooling schedules, such as the geometric or the linear ones [2], make SA nonconvergent: the Monte Carlo walker has a nonzero probability of getting trapped into minima other than the global one.

The cooling schedule implied by Equation 6.2 is, of course, asymptotically valid in the limit of low temperatures. In the same limit, PTMC allows for a geometric temperature schedule [30,31]. When the temperature drops to zero, the system is well approximated by a multidimensional harmonic oscillator and the acceptance probability for swaps attempted between two replicas with temperatures $T < T'$ is given by the incomplete beta function law [31]

$$\mathrm{Ac}(T, T') \simeq \frac{2}{B(d/2, d/2)} \int_0^{1/(1+R)} \theta^{d/2-1}(1-\theta)^{d/2-1} \, d\theta, \tag{6.3}$$

where d denotes the number of degrees of freedom of the system, B is the Euler beta function, and $R \equiv T'/T$. Since it depends only on the temperature ratio R, the acceptance probability (6.3) has the same value for any arbitrary replica running at a temperature T_i, provided that its neighboring upper temperature T_{i+1} is given by $T_{i+1} = RT_i$. The value of R is determined such that the acceptance probability given by Equation 6.3 attains a prescribed value p, usually chosen greater that 0.5. Thus, the (optimal) schedule that ensures a constant probability p for swaps between neighboring temperatures is a geometric progression:

$$T_i = R^{i-1} T_{\min}, \quad 1 \leq i \leq N, \tag{6.4}$$

where $T_{\min} = T_1$ is the minimum temperature of the schedule. Although more research is required to assess the relative efficiency of the two different algorithms, it is apparent from Equations 6.2 and 6.4 that PTMC is a global optimizer superior to SA because it allows for a faster cooling schedule. Direct numerical comparisons of the two methods have confirmed that parallel tempering is the superior optimization technique [32]. The ideas of parallel tempering and simulated annealing are not mutually exclusive, and in fact they can be used together to design even more efficient stochastic optimizers. As shown below, such a strategy that combines parallel tempering and simulated annealing is employed for the present simulations.

6.1.3
Description of the Algorithm

The typical Monte Carlo simulation done in this work consists of two main parts that are equal in terms of computational effort. In the first stage of the computation, we perform a parallel tempering run for a range of temperatures $[T_{\min}, T_{\max}]$. The configurations of minimum energy are retained for each replica, and used as starting configurations for the second part of the simulation, in which each replica is cooled down exponentially until the largest temperature drops below a prescribed value. As a key feature of the procedure, the parallel tempering swaps are not turned off during the cooling stage. Thus, we are using a combination of parallel tempering and simulated annealing, rather than a simple cooling. Finally, the annealed replicas are relaxed to the nearest minima using a conjugate gradient algorithm. We now describe in detail the stochastic minimization procedure. We shall focus, in turn, on discussing the Monte Carlo moves, the choice of the temperature range $[T_{\min}, T_{\max}]$, and the total number of necessary replicas N.

The moves of the hot atoms consist in small random displacements with the x, y, z components given by

$$\Delta(2u_\alpha - 1),$$

where $u_\alpha (\alpha = x, y, z)$ are independent random variables [33] uniformly distributed in the interval $[0, 1]$, and Δ is the maximum absolute value of the displacement. We update the positions of the hot atoms one at a time in a cyclic fashion. Each attempted move is accepted or rejected according to the Metropolis logic [26]. A complete cycle consisting in attempted moves for all hot particles is called a *pass* (or sweep) and constitutes the basic computational unit in this work. We have computed distinct acceptance probabilities for the hot atoms that are closer to the surface (within 5 Å) and for the deeper atoms, the movements of which are essentially small oscillations around the equilibrium bulk positions. Consequently, we have employed two different maximal displacements, Δ_s for the surface atoms and Δ_b for the bulk-like atoms lying in the deeper layers. The displacements Δ_s and Δ_b are tuned automatically in the equilibration phase of the simulation in such a way that the Monte Carlo moves are accepted with a rate of 40–60%. Global moves (i.e., configuration swaps) are attempted between replicas running at neighboring temperatures at every 10 passes in an alternating manner, first with the closest lower temperature and then with the closest higher temperature. Exception is the two replicas that run at end temperatures $T_1 = T_{\min}$ and $T_N = T_{\max}$, which are involved in swaps every 20 passes. The range of temperatures $[T_{\min}, T_{\max}]$ and the temperature schedule $T_1 < T_2 < \cdots < T_N$ have been chosen as described below.

The maximum temperature $T_{\max}$ must be high enough to ensure that the corresponding random walker has good probability of escaping from various local minima. Therefore, we set the high temperature equal to the melting temperature of the surface slab, which can be determined from a separate PTMC simulation by identifying the peak of the heat capacity as a function of temperature. As Figure 6.1 shows, the melting

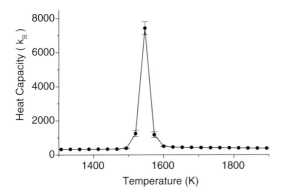

Figure 6.1 Heat capacity of a Si(105) slab plotted as a function of temperature. The peak is located at 1550 K; in order to avoid recalculation of the heat capacity for systems with different numbers of atoms and surface orientations, we set $T_{\max} = 1600$ K as the upper limit of the temperatures range used in the PTMC simulations. (From Ref. [44], with permission from American Physical Society.)

temperature of a Si(105) sample slab with 70 hot atoms is about 1550 K. Rather than determining a melting temperature for each individual system studied, we have employed a fixed value of $T_{max} = 1600$ K. The melting temperature of the slab determined here (Figure 6.1) is different from the value of 1250 K reported for the bulk crystal [25]: the discrepancy is due to surface effects and finite-size effects, as well as due to the fact that the hot atoms are always in contact with the rigid atoms from the bottom of the slab. Though we use $T_{max} = 1600$ K for all simulations, we note that differences of 100–200 K in the melting temperature of the slab do not significantly affect the quality of the Monte Carlo sampling. For most surfaces and system sizes of practical importance, the value of 1600 K is in fact an upper bound for the melting temperature; this may sometimes cause the one or two walkers that run at the highest temperatures to be uncoupled from the rest of the simulation, since they might sample amorphous or liquid states. However, this loss in computational resources is very small compared to the additional effort that would be required by a separate determination of the heat capacity for each surface slab used.

In theory, the lowest temperature T_{min} should be set so low that the walker associated with this temperature is virtually localized in the basin associated with the global minima. Nevertheless, obstacles concerning the efficient use of computational resources prevent us from doing so. Numerical experimentation has shown that a temperature of $T_{min} = 400$ K is low enough that only local minima associated with realistic surface reconstructions are frequently visited. A further selection among these local minima is performed in the second part of the Monte Carlo simulation, when all temperatures of the initial schedule $\{T_i, i = 1, 2, \ldots, N\}$ are gradually lowered to values below 100 K; as it turns out, this combination of parallel tempering and simulated annealing makes optimal use of computational resources. Below the melting point, the heat capacity of the surface slab is almost constant and well approximated by the capacity of a multidimensional harmonic oscillator (refer to Figure 6.1). Under these conditions, the acceptance probability for swaps between neighboring temperatures T and T' is given by Equation 6.3 (see also Ref. [31]). It follows that the optimal temperature schedule in the interval $[T_{min}, T_{max}]$ is the geometric progression (6.4), where

$$R = (T_{max}/T_{min})^{1/[N(d,p)-1]}.$$

We have written $N \equiv N(d, p)$ to denote the smallest number of replicas that guarantees a swap acceptance probability of at least p for a system with d degrees of freedom. Since the best way to run PTMC calculations is to use one processor for each replica of the system, the feasibility of our simulations hinges on values of $N(d, p)$ that translate directly into available processors. The number of walkers $N(d, p)$ can be estimated [31] by

$$N(d, p) = \left[d^{1/2} \frac{\sqrt{2}\ln(T_{max}/T_{min})}{4\mathrm{erf}^{-1}(1-p)} \right] + 2, \qquad (6.5)$$

where $[x]$ denotes the largest integer smaller than x and erf^{-1} is the inverse error function. Based on Equation 6.5, we have used $N = 32$ walkers for all simulations, which ensures a swap acceptance ratio greater than $p = 0.5$ for any system with less than 300 hot atoms, $d < 900$. The first part of all Monte Carlo simulations performed in this chapter consists of a number of 36×10^4 passes for each replica, preceded by 9×10^4 passes allowed for the equilibration phase. When we retained the configurations of minimum energy, the equilibration passes have been discarded so that any memory of the starting configuration is lost.

We now describe the second part of the Monte Carlo simulation, which consists of a combination of simulated annealing and parallel tempering. At the kth cooling step, each temperature from the initial temperature schedule $\{T_i, i = 1, 2, \ldots, N\}$ is decreased by a factor that is independent of the index i of the replica, $T_i^{(k)} = \alpha_k T_i^{(k-1)}$. Because the parallel tempering swaps are not turned off, we require that at any cooling step k all N temperatures must be modified by the same factor α_k in order to preserve the original swap acceptance probabilities. The specific way in which α_k depends on the cooling step index k is determined by the kind of annealing being sought. In this work, we have used a cooling schedule of the form

$$T_i^{(k)} = \alpha T_i^{(k-1)} = \alpha^{k-1} T_i \quad (k \geq 1), \tag{6.6}$$

where $T_i \equiv T_i^{(1)}$ and α is determined such that the temperature intervals $[T_1^{(k-1)}, T_N^{(k-1)}]$ and $[T_1^{(k)}, T_N^{(k)}]$ spanned by the parallel tempering schedules before and after the kth cooling step overlap by 80%. This yields a value for α given by $(0.2 T_{\min} + 0.8 T_{\max})/T_{\max} = 0.85$.

The reader may argue that the use of an exponential annealing (Equation 6.6) is not the best option for attaining the global energy minimum of the system. Apart from the theoretical considerations discussed in the preceding subsection that only a logarithmic cooling schedule would ensure convergence to the ground state [28,29], it is known that the best annealing schedules for a given computational effort oftentimes involve several cooling–heating cycles. We emphasize that in the present simulations, the most difficult part of the sampling is taken care of by the initial PTMC run. In addition, since the configuration swaps are not turned off during cooling (refer to Figure 6.2), the Monte Carlo walkers are subjected to cooling–heating cycles through the parallel tempering algorithm.

The purpose of the annealing (second part of the simulation) is to cool down the best configurations determined by the initial parallel tempering in a way that is more robust than the mere relaxation into the nearest local minimum. If the initial PTMC run is responsible for placing the system in the correct funnels (groups of local minima separated by very large energy barriers), the annealing part of the simulation takes care of jumps between local minima separated by small barriers within a certain funnel. For this reason, the annealing is started from the configurations of minimum energy determined during the first part. The cooling is stopped when the largest temperature in the parallel tempering schedule drops below 100 K. This criterion yields a total of 18 cooling steps, with 2×10^4 MC

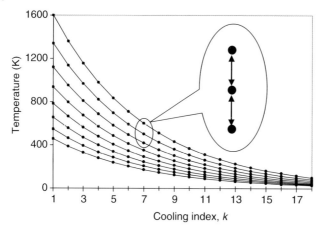

Figure 6.2 Exponential cooling of the $N = 32$ Monte Carlo walkers (replicas of the surface slab) used in the simulation. For clarity, only eight walkers are shown (every fourth walker). The cooling is performed in 18 steps: at each step, the temperature is modified by the same factor $\alpha = 0.85$ for all walkers (Equation 6.6). For every cooling step k, we have a different parallel tempering schedule where each replica is coupled to the walkers running at neighboring temperatures via configuration swaps (Equation 6.4 with $R = 4^{1/31}$). This coupling is symbolized by the double-arrow lines in the inset. (From Ref. [44], with permission from American Physical Society.)

passes per replica performed at every such step. Each cooling step is preceded by 5×10^3 equilibration passes, which are also used for the calculation of new maximal displacements Δ_s and Δ_b, as these displacements depend on temperature and must be recomputed. In fact, each cooling step is a small-scale version of the first part of the simulation. The only difference is that the cooling steps are *not* started from the configurations of minimum energy determined at the preceding cooling steps.

The third and final part of the minimization procedure is a conjugate gradient optimization of the last configurations attained by each replica. The relaxation is necessary because we aim to classify the reconstructions in a way that does not depend on temperature, so we compute the surface energy γ at zero kelvin for the relaxed slabs i, $i = 1, 2, \ldots, N$.

6.2
Basin Hopping Monte Carlo

Few methods enjoy the level of ingenuity, power, and simplicity that GA does, and BHMC is one of them. In BHMC, we do not deal with the PES, but rather with the reduced PES that is obtained by local relaxation from any given starting point. Therefore, the energy at any point in configuration space is prescribed to be that of the local minimum obtained by the given geometry optimization technique: As

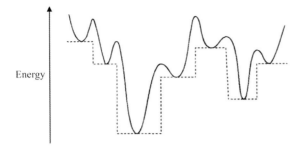

Figure 6.3 Schematic sequence of staircases and plateaus (basins) obtained by local relaxation of the potential energy surface at every point in the configuration space. (From Ref. [6], with permission from American Chemical Society.)

Doye and Wales describe in their original work [6], the "PES is mapped onto a set of interpenetrating staircases with plateaus corresponding to the set of configurations which lead to a given minimum after optimization."

This is illustrated in Figure 6.3. These plateaus are also called basins of attraction, and have been useful as a means to compare the efficiency of different transition state searching techniques. The *reduced energy landscape* for a given system was then sampled using a canonical Monte Carlo simulation at a constant reduced temperature. In other words, the sampling technique is the same as one of the Monte Carlo walkers described using simulated annealing (see the previous section); only these walkers are used to explore a simpler, reduced PES! Again similar to the SA and PTMC, the maximal displacement of any of the spatial coordinates is adjusted dynamically such that the acceptance ratio is close to 0.5. Note that in the case of BHMC, the nature of the reduced PES coupled with the dynamical adjustment of maximal displacement results in relatively large step sizes compared to PTMC. There are certain cluster sizes for which the global minimum is not based on an icosahedral structure, and were particularly challenging to obtain prior to the introduction of BHMC; these structures (depicted in Figure 6.4) were readily

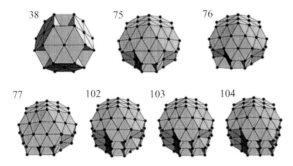

Figure 6.4 Nonicosahedral global minima of Lennard-Jones clusters obtained via basin hopping Monte Carlo. (From Ref. [6], with permission from American Chemical Society.)

recovered by BHMC as the first of many of its successes in optimization of atomic clusters.

6.3
Optimization via Minima Hopping

With the exception of GA, a wide variety of structure optimization algorithms (see, for example, SA, PTMC, and BHMC above) function based on thermodynamics since a Markov process (Metropolis algorithm [26]) with a Boltzmann factor leads to the thermodynamic distribution. In principle, at sufficiently low temperature the ground-state configuration will emerge as the predominant structure and the problem is apparently solved – but, as mentioned, this takes a very long time, and it may even be intractable computationally unless temperature scheduling schemes are employed. This is why the prevailing "wisdom" is that thermodynamics should not be used for finding global minima.

As we have seen throughout this book, GA tries to explore as much of the configuration space as possible, and as efficiently as possible. Minima hopping aims for the same thing. However, visiting the same structures or configurations over and over again slows the method down – in fact, all Monte Carlo methods have this exact tendency of revisiting configurations over and over again. A proposed method to avoid revisits was flooding: the basins, or local minima, or funnels that were already visited during the simulation are elevated (artificially) so that it becomes very hard for them to be visited again. However simple, this approach never caught on because it is extraordinarily complicated to determine the boundaries of the region that should be elevated. The second problem with flooding is that sometimes an intermediate or transitional basin is flooded and that (often) leads to blocking access to the ground-state funnel: so if the simulation started in an unfortunate place, and flooding is applied, it may never get to the ground state. Based on these arguments, Goedecker concludes that what is necessary is a technique that discourages (which not strictly forbids) repeated visits of local minima without blocking access to important funnels or basins; he dubbed this the MH method [7].

The MH algorithm in summarized in the flowchart of Figure 6.5 and it consists of two major loops. We follow here the original explanation of the principle of MH minimization by Goedecker [7], and use the same notations in order to avoid confusion. The inner loop executes jumps into the different local minima, while the outer loop accepts or rejects such jumps: this is done based on a threshold, such that the jump or step is accepted if the energy of the new minimum increases less than E_{diff} compared to the energy of the present minimum. The parameter E_{diff} is dynamically changed so that only half of the moves are accepted and the rest rejected. The outer loop biases for downward moves in energy.

```
MDstart
  ESCAPE TRIAL PART
    start a MD trajectory with kinetic energy Ekinetic
    from current minimum 'Mcurrent'. Once the
    potential reaches the mdmin-th minimum
    along the trajectory stop MD and optimize
    geometry to find the closest local minimum 'M'
    if ('M' equals 'Mcurrent') then
      Ekinetic=Ekinetic*beta1 (beta1>1)
      goto MDstart
    else if ('M' equals a minimum visited previously)
    then
      Ekinetic=Ekinetic*beta2 (beta2>1)
      goto MDstart
    else if ('M' equals a new minimum) then
      Ekinetic=Ekinetic*beta3 (beta3<1)
    endif
  DOWNWARD PREFERENCE PART
    if (energy('M')-energy('Mcurrent')<Ediff) then
      accept new minimum: 'Mcurrent'='M'
      add 'Mcurrent' to history list
      Ediff=Ediff*alpha1 (alpha1<1)
    else if rejected
      Ediff=Ediff*alpha2 (alpha2>1)
    endif
    goto MDstart
```

Figure 6.5 Explicit flowchart of the MH algorithm. (From Ref. [7], with permission from American Institute of Physics.)

The inner part consists of the so-called escape step out of the present local minimum followed by a geometry relaxation into the possibly other local minimum. The escape step is a short molecular dynamics run in which the atoms have a Boltzmann velocity distribution such that their kinetic energy is equal to $E_{kinetic}$. If $E_{kinetic}$ is small, then the system falls back into the present well. Otherwise, the system will most likely jump away from the present well (minimum) and reach a different one. The molecular dynamics simulation is stopped as soon as the potential energy has crossed a prescribed number of maxima and reached the minimum along the molecular dynamics trajectory. After the molecular dynamics is stopped, a local relaxation is executed.

Three different cases can occur as a result of the jumps in the inner loop. (A) In the first case, the relaxation can yield the present minimum. (B) In the second case, the new well (minimum) is one that has already been visited before during the simulation. (C) In the third case, the minimum is truly a new one that has not been visited before: this outcome is the one that leads to progress in the MH method, since it helps in exploring previously unexplored configurations and eventually leads to the global minimum.

Since we want to explore new configurations, the first temptation is to increase $E_{kinetic}$. However, increasing it indiscriminately makes the system cross high-energy barrier: here lies exactly the problem, because crossing high barrier leads the systems into minima with similarly high energies (refer to Figure 6.6). Such correlation is called the Bell–Evans–Polanyi principle, and recognizing it was a

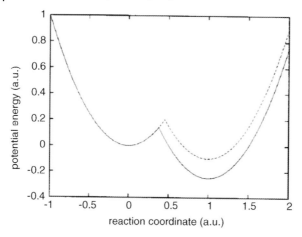

Figure 6.6 Illustration of the Bell–Evans–Polanyi principle. The transition state is raised if the funnel on the right-hand side is raised. (From Ref. [7], with permission from American Institute of Physics.)

key factor for the success of the MH method. On the other hand, making $E_{kinetic}$ very small means that we should start many molecular dynamics runs in order to have a reasonable chance to escape the current minimum and explore others. Therefore, in practice it turns out to allow $E_{kinetic}$ to be adjusted automatically (dynamically) during the run such that half of the molecular dynamics simulations end up taking the system away from the current well into another one. The value of $E_{kinetic}$ is increased dynamically not only if one falls back into the old well but also if the system visits wells that it has already visited before. Thus, in order to decide whether this parameter has to be increased or not, one has to track and catalogue all the minima that were already visited.

The flowchart of the MH shows its five parameters (Figure 6.5). α_1 and α_2 dictate how quickly E_{diff} is increased or decreased when a new configuration is rejected or accepted. β_1, β_2, and β_3 determine how quickly $E_{kinetic}$ changes following the outcome of an escape molecular dynamics move. With increased values the global minimum is found faster in the cases where it is found, but there are also cases where it can be missed. Since $E_{kinetic}$ is increased after visiting previously visited states, it follows that after the system has explored the low-energy configurations it starts to explore higher and higher regions on the PES. At some point, simulation is stopped when sufficiently higher regions have been explored (e.g., a desired energy range above the lowest energy).

Finally, we note that the MH method has recently been modified for periodic structures in three dimensions and now represents a very powerful crystal prediction method [34]. Known successes are the diamond structure of silicon and the structure of AB Lennard-Jones binary alloys (Figure 6.7). Interestingly, MH has been applied to rather diverse problems such as the structures of scanning tunneling microscopy (STM) tips [35], the problem of protein folding [36], and the determination of low-energy defects [37].

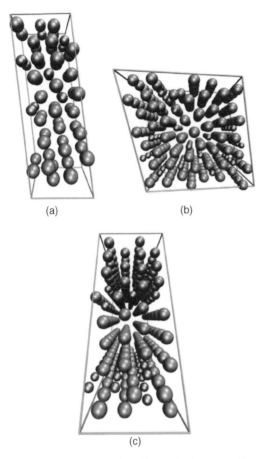

Figure 6.7 Structure predicted by MH for AB Lennard-Jones mixtures with 60 (a), 256 (b), and 320 (c) atoms. Type A atoms are denoted by red (large) spheres and type B atoms are blue (small). (From Ref. [34], with permission from American Institute of Physics.)

6.4
The Metadynamics Approach

Predicting equilibrium crystal structures for given *pressure* and temperature conditions is a key question in nanoscale and condensed matter physics, materials science, geophysics, planetary physics, polymer science, and others. Most crystals exhibit phase transitions when an external pressure is applied and increased, but the final structure is usually unknown and almost always difficult to identify, so computer simulations become necessary in identifying these final structures. Progress has been made when the constant-pressure molecular dynamics was introduced [38], and, in particular, the Parrinello–Rahman technique [39]. In the Parrinello–Rahman technique, the simulation cell is allowed to change its shape and volume in order to comply with the structure toward which applied pressure tends to drive the materials system. The Parrinello–Rahman method now appears in textbooks and is widely used in different versions.

However, we have to realize that phase transformations are often of the first order, and with the use of periodic boundary conditions and small unit cells, the heterogeneous nucleation of the new phase is largely suppressed and one has to overpressurize the system in order to induce the transformation. With overpressurization, the dynamical pathway may not always be realistic and one may end up in a different and artificial new state rather than an experimentally relevant one. Martoňák et al. [40] have devised a method to overcome these problems. This is based on the realization that the optical phonons (atomic vibrations) are much faster than the acoustical ones (vibrations of the crystalline dimensions) and therefore are well separated in timescales. One can therefore use the cell dimensions and angles as order parameters and design simple differential equations that govern the evolution of these order parameters: this evolution of the metadynamics is the key to simulating the phase transformation [40]. At each metadynamics step, we still have molecular dynamics simulations in play that are necessary for moving along the fast degrees of freedom. We mention here that, as an early success of this method, the diamond structure of bulk silicon was seen to change to hexagonal form with applied pressure (Figure 6.8).

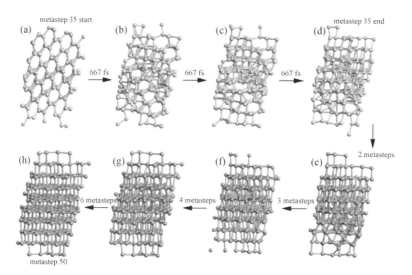

Figure 6.8 (a–d) Evolution of atomic configurations during 2 ps of microscopic dynamics (at intervals of 667 fs) across the transition at metastep 35. The initial diamond structure (a) is strongly strained, compressed along one axis, and elongated along perpendicular ones. In (b) and (c), the diamond structure makes room for a new periodic structure (d). (e–h) Evolution during 15 subsequent steps of metadynamics. Note the gradual formation of the simple hexagonal phase. (From Ref. [40], with permission from American Physical Society.)

6.5
Comparative Studies between GA and Other Structural Optimization Techniques

At present, not too many comparative studies between different global optimization methods are being published. In the following, we describe two such comparisons from our own work. The first one compares GA and PTMC for surface reconstructions [41], and the second one compares GA and MH for the crystal structure prediction problem [42]. Some of the broad messages resulting from these comparisons will also be repeated in Chapter 7.

6.5.1
Reconstructions of Si(114): Comparison between GA and PTMC

In this section, we use PTMC and GA to report the results of global structural optimization of the Si(114) surface, a stable high-index surface orientation of silicon. These optimization procedures, coupled with the use of a highly optimized interatomic potential for silicon, lead to finding a set of possible models for Si(114), whose energies are recalculated with *ab initio* density functional methods. The most stable structure obtained here without experimental input coincides with the structure determined from scanning tunneling microscopy experiments and density functional calculations by Erwin, Baski, and Whitman [43].

Treating the reconstruction of semiconductor surfaces as a problem of global optimization, Ciobanu and Predescu have developed a parallel tempering Monte Carlo procedure for studying the structure and thermodynamics of crystal surfaces [44], which we compare here with GA [45]. The use of such methods can help avoid situations in which the actual physical reconstruction of a high-index surface is not part of the set of heuristic models that are considered for computation of surface energies and comparison with experimental data. Given that there are examples of semiconductor surfaces (e.g., see Refs [46,47]) for which the initially proposed models did not withstand further scientific scrutiny from different research groups, it is worthwhile to perform searches for the structure of some of stable high-index surface orientations of silicon. One such surface is Si(114), reported to be as stable as the well-studied low-index surfaces Si(001) and Si(111) [43]: given this stability of Si(114), it is somewhat surprising that this surface has not attracted more interest, at least from a technological perspective.

Based on STM images combined with density functional calculations, two atomic models [(2×1) and $c(2 \times 2)$] were proposed for the Si(114) orientation [43]. These models have very similar bonding topology, differing only in terms of dimerization pattern of their surface. The surface energies of the two models are also similar, as both can be found on sufficiently large areas of the scanned samples [43]. To our knowledge, so far Ref. [43] represents the *only* proposal for the structure of Si(114). The purpose of this chapter is to present several other model candidates for the structure of Si(114), models that are likely to be experimentally observed on this surface. Addressing the problem of atomic structure from a different perspective than the previous reports, we perform stochastic searches for the global minimum configuration of this surface. As we shall see, the lowest-energy configuration (at zero kelvin) obtained here from purely theoretical means is consistent with the original proposal [43]; however, the global search methods provide several other structural models with low surface energy, which could be relevant in various experimental conditions.

6.5.1.1 PTMC Results

At the end of the PTMC simulation (Section 6.1), we analyze the energies of the relaxed replicas. Typical plots showing the surface energies of the structures retrieved by the PTMC replicas are shown in Figure 6.9a, for different numbers of particles in the computational cell. To exhaust all the possibilities for the numbers

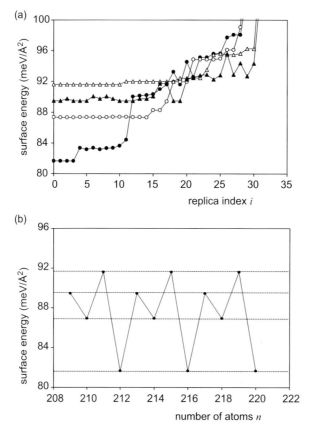

Figure 6.9 (a) Surface energies of the relaxed parallel tempering replicas $i(0 \leq i \leq 31)$, with total number of atoms $n = 216$ (solid circles), 215 (open triangles), 214 (open circles), and 213 (solid triangles). For clarity, the range of plotted surface energies was limited from above at 100 meV/Å^2. (b) Surface energy of the global minimum structure showing a periodic behavior as a function of n, with a period of $\Delta n = 4$; this finding helps narrowing down the set of values for n that need to be considered for determining the Si(114) reconstructions that have a $3a \times a\sqrt{2}$ periodic cell. (From Ref. [41], with permission from Elsevier.)

of particles corresponding to the supercell dimensions of $3a \times a\sqrt{2}$, we repeat the PTMC simulation for different values of n ranging from 208 to 220, and look for a periodic behavior of the lowest surface energy as a function of n. For the case of Si(114), this periodicity occurs at intervals of $\Delta n = 4$, as shown in Figure 6.9b. Therefore, the (correct) number of atoms n at which the lowest surface energy is attained is $n = 216$, up to an integer multiple of Δn. As we shall show in the next section, the repetition of the simulation for different values of n in the simulation cell can be avoided within a genetic algorithm approach.

6.5.1.2 GA Results

As described in Chapter 5, we have developed two versions of the algorithm. In the first version, the number of atoms n is kept the same for every member of the pool by automatically rejecting child structures that have different numbers of atoms from their parents (mutants). In the second version of the algorithm, this restriction is not enforced, that is, mutants are allowed to be part of the pool: in this case, the procedure is able to select the correct number of atoms for the ground-state reconstruction without any increase in the computational effort required for one single constant-n run. The results of a variable-n run are shown in Figure 6.10a, which shows how the lowest energy and the average energy from a pool of $p = 30$ structures decrease as the genetic algorithm run proceeds. The plot in Figure 6.10a displays typical features of the evolutionary approach: the most unfavorable structures are eliminated from the pool rather fast (initial steep transient region of the graphs) and a longer time is taken for the algorithm to retrieve the most stable configuration. The lowest-energy structure is retrieved in less than 500 mating operations. The correct number of atoms (refer to Figure 6.10b) is retrieved much faster, within approximately 100 operations. It is worth noting that even during the transient period, the lowest-energy member of the pool spends most of its evolution in a state with a number of atoms ($n = 212$) that is compatible with the global minimum structure.

The two independent algorithms, PTMC and GA, presented briefly in this section are able to retrieve a set of possible candidates for the lowest-energy surface structure. We use both the algorithms in this work in order to assess how robust their structure predictions are. As it turns out, the two methods not only find the same lowest-energy structures for each value of the total number of atoms n, but also find most of the other low-energy reconstructions – a finding that builds confidence in the quality of the configuration sampling performed here. Since the atomic interactions are modeled by an empirical potential [25], it is desirable to check the relative stability of different model structures using higher-level calculations based on density functional theory (DFT); the details of these calculations are presented next.

6.5.1.3 DFT Calculations

Using the methodologies described above, we build a database of model structures that are sorted according to the surface energy given by the HOEP [25] interaction model. Since the empirical potentials may not give a reliable energetic ordering when a large number of structures are considered, we recalculate the surface

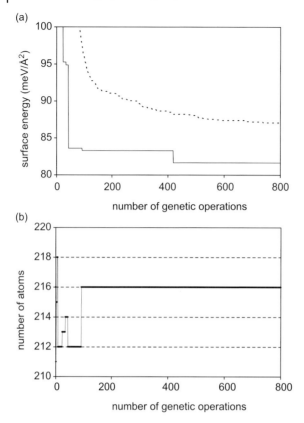

Figure 6.10 (a) Evolution of the lowest surface energy (solid line) and the average energy (dashed line) for a pool of $p = 30$ structures during a GA run with variable $n(210 \leq n \leq 222)$. (b) Evolution of the number of atoms n that corresponds to the model with the lowest energy from the pool, during the same GA run. Note that the lowest-energy structure of the pool spends most of its evolution in states with numbers of atoms that are compatible with the global minimum, that is, $n = 212$ and $n = 216$. (From Ref. [41], with permission from Elsevier.)

energies of the models in the database at the level of density functional theory. The calculations were performed with the plane-wave-based PWscf package [48], using the Perdew–Zunger [49] exchange-correlation energy. The slab geometry and the computational parameters are similar to the ones reported in Ref. [43]; given the increase in computational speed over the past 8 years, we used thicker slabs and a different sampling of the Brillouin zone. The cutoff for the plane-wave energy was set to 12 Ry, and the irreducible Brillouin zone was sampled with four k-points. The equilibrium bulk lattice constant was determined to be $a = 5.41$ Å, which was used for all the surface calculations in this work. The simulation cell has the single-face slab geometry, with 24 layers of Si, and a vacuum thickness of 12 Å. The bottom three layers are kept fixed in order to simulate the underlying bulk geometry, and the lowest layer is passivated with hydrogen. The remaining Si layers are allowed to relax until the forces become smaller than 0.025 eV/Å.

The surface energy γ for each reconstruction is determined indirectly, by first considering the surface energy γ_0 of an unrelaxed bulk truncated slab, and then calculating the difference $\Delta\gamma = \gamma - \gamma_0$ between the surface energy of the actual reconstruction and the surface energy of a bulk truncated slab that has the bottom three layers fixed and hydrogenated. The energy of the bulk truncated surface, as computed from a two-faced slab with 24 layers, was found to be $\gamma_0 = 143\,\text{meV}/\text{Å}^2$.

6.5.1.4 Structural Models for Si(114)

At the end of the global search procedures, we obtain a set of model structures that we sort by the number of atoms in the simulation cell and by their surface energy. Since the empirical potentials may not be fully transferable to different surface environments, we study not only the global minima given by the model for different values of n, but also most of the local minima that are within $15\,\text{meV}/\text{Å}^2$ from the lowest-energy configurations. After the global optimizations, the structures obtained are also relaxed by DFT methods as described in Section 6.5.1.3. The results are summarized in Table 6.1, which will be discussed next.

Table 6.1 lists the density of dangling bonds (dbs per area) and the surface energies of several different models calculated using the HOEP potential and DFT.

Table 6.1 Surface energies of different reconstructions for the Si(114) surface, sorted by the number of atoms n in the $3a \times a\sqrt{2}$ periodic cell.

n	Bond counting (db/$3a^2\sqrt{2}$)	HOEP (meV/Å^2)	DFT (meV/Å^2)
216	8	81.66	89.48
	8	83.16	90.34
	8	83.31	91.29
	8	83.39	88.77
	8	83.64	94.68
	8	84.42	92.16
215	8	91.61	97.53
	8	91.82	95.30
	8	92.00	94.20
	11	92.46	98.73
214	6	86.95	95.17
	10	87.32	99.58
	10	87.39	98.47
	10	87.49	93.88
	10	88.26	95.18
213	4	89.46	90.43
	6	89.76	94.01
	4	90.07	90.85
	6	91.73	94.66
	7	93.99	90.48

The second column shows the number of dangling bonds (counted for structures relaxed with HOEP) per unit area. The last two columns list the surface energies given by the HOEP interaction model [25] and by density functional calculations [48] with the parameters described in the text.

The configurations have been listed in increasing order of the surface energies computed with HOEP, as this is the actual outcome of the global optimum searches. For reasons of space, we limit the number of structures in Table 6.1 to at most six for each value of the relevant numbers of atoms in the simulation cell. However, when performing DFT relaxations we consider more structures than the ones shown in the table because we expect changes in their energetic ordering at the DFT level. The inclusion of a larger number of structures helps avoid excessive reliance on the empirical potential [25], which is mainly used as a fast way to provide physically relevant reconstructions (i.e., where each atom at the surface has at most one dangling bond).

Table 6.1 and Figure 6.9b suggest that the most unfavorable number of atoms in the simulation cell is $n = 215$, both at the level of HOEP and at the level of DFT. Therefore, it is justifiable to focus our attention on the other three values of n ($n = 216, 214$, and 213), which yield considerably lower surface energies. For each of these numbers of atoms, we present four low-energy structures (as given by DFT), which are shown in Figures 6.11–6.13. These structures are not necessarily the same as those enumerated in Table 6.1, as they are chosen based on their DFT surface energies. Since the global optimization has not been performed at the DFT level, the reader could argue that the lowest-energy structure obtained after the sorting of the DFT-relaxed models may not be the DFT global minimum. While we found that a *thorough* sampling for systems with $\sim$200 atoms is impractical at the DFT level, we have performed DFT relaxations for most of the local minima given by HOEP. Therefore, given the rather large set of structural candidates with different topological features considered here, the possibility of missing the actual reconstruction for Si(114) is much diminished in comparison with heuristic approaches.

We will now describe in turn the surface models corresponding to $n = 216, 214$, and 213. After the DFT relaxation, the lowest-energy model that we found has turned out to be the same as the one proposed by Erwin *et al.* [43], perhaps with the exception of different relative tilting of the surface bonds. The model is shown in Figure 6.11a, and it is characterized by the presence of dimers, rebonded atoms, and tetramers occurring in this order along the (positive) [$22\bar{1}$] direction. These features have been well studied [43], and we shall not insist on them here. The surface energy of the most stable model for Si(114)-(2 × 1) reconstruction is $\gamma = 88.77 \, \mathrm{meV/\AA^2}$. Although this surface energy is somewhat different from the previously reported value of $85 \, \mathrm{meV/\AA^2}$ [43], the discrepancy between these absolute values can be attributed to the somewhat different computational parameters (slab thickness, number of k points) and different pseudopotentials.

It is notable that a different succession of the above-mentioned atomic-scale features is also characterized by a low surface energy: specifically, dimers, tetramers, and rebonded atoms (in this order along [$22\bar{1}$]), as shown in Figure 6.11c, give a surface energy that is only $\sim$2 meV/$\mathrm{\AA}^2$ higher than that of the Erwin *et al.*'s model shown in Figure 6.11a. This surface energy gap is apparently large enough to allow for another configuration (see Figure 6.11b) with a surface energy that lies between the values corresponding to the first two models described above. As shown in Figure 6.11b, this new model has two consecutive dimer rows followed by a row of

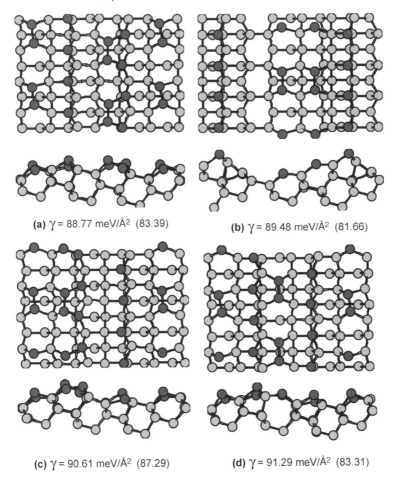

Figure 6.11 Structural models (top and side views) of Si(114)-2 × 1, with $n = 216$ atoms per unit cell after relaxation with density functional methods [48]. The surface energy γ computed from first-principles calculations is indicated for each structure, along with the corresponding value (in parentheses) obtained using the empirical potential [25]. The darker shade marks the undercoordinated atoms. (From Ref. [41], with permission from Elsevier.)

rebonded atoms, and arrangement that gives rise to surface corrugations of 0.4–0.5 nm. Remarkably, this corrugated model (Figure 6.11b) is almost degenerate with the planar (2 × 1) structure shown in Figure 6.11a. Figure 6.11d shows another planar model of Si(114), made of dimers, rebonded atoms, and inverted tetramers, with the latter topological feature distinguishable as a seven-member ring when viewed along the [110] direction.

Dimers, rebonded atoms, and tetramers also occur on low-energy structural models with $n = 214$, as shown in Figure 6.12. The most favorable structure with $n = 214$ that we found (depicted in Figure 6.12a) has a five-coordinated subsurface atom and a four-coordinated surface atom per unit cell. These topological features

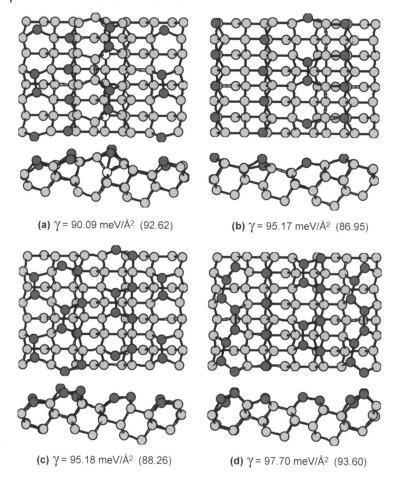

Figure 6.12 Structural models (top and side views) of Si(114)-2 × 1, with $n = 214$ atoms per unit cell after relaxation with density functional methods [48]. The surface energy γ computed from first-principles calculations is indicated for each structure, along with the corresponding value (in parentheses) obtained using the empirical potential [25]. The darker shade marks the undercoordinated atoms, while the overcoordinated atoms are shown in white. (From Ref. [41], with permission from Elsevier.)

are determined by the bonding of a subsurface atom with one of the atoms of a tilted surface dimer; the corresponding surface energy is 90.09 meV/Å². Other structures with $n = 214$ atoms (examples shown in Figure 6.12b–d) generally have higher energies than models with $n = 216$ (refer to Table 6.1), most likely because the two missing atoms lead to pronounced strains in the surface bonds.

The analysis of simulation slabs with $n = 213$ atoms reveals novel atomic-scale features. Energetically favorable configurations with $n = 213$ (Figure 6.13a and c) show a five-atom ring on the surface stabilized by a subsurface interstitial, a structural complex that was first encountered in the case of Si(113) surface [50].

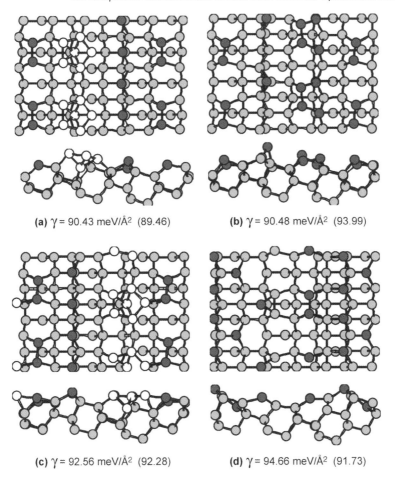

(a) $\gamma = 90.43$ meV/Å² (89.46)

(b) $\gamma = 90.48$ meV/Å² (93.99)

(c) $\gamma = 92.56$ meV/Å² (92.28)

(d) $\gamma = 94.66$ meV/Å² (91.73)

Figure 6.13 Structural models (top and side views) of Si(114)-2 × 1, with $n = 213$ atoms per unit cell after relaxation with density functional methods [48]. The surface energy γ computed from first-principles calculations is indicated for each structure, along with the corresponding value (in parentheses) obtained using the empirical potential [25]. The darker shade marks the undercoordinated atoms, while the white atoms are either four-coordinated or overcoordinated. (From Ref. [41], with permission from Elsevier.)

Structures in Figure 6.13a and c differ in terms of the succession of the topological features along the [22$\bar{1}$] direction, that is, dimers, five-member rings, and rebonded atoms (Figure 6.13a) as opposed to dimers, rebonded atoms, and five-member rings (Figure 6.13c). The model in Figure 6.13a is degenerate with the one shown in Figure 6.13b, as their relative surface energy is much smaller than the 1–2 meV/Å² expected accuracy of the relative surface energies determined here. The reconstruction in Figure 6.13b is very similar to the lowest-energy structure in Figure 6.11a (achievable with $n = 212$): the only different feature is the extra atom lying in between two rebonded atoms and sticking out of the surface (refer to

Figure 6.13b). Likewise, the model in Figure 6.13d can be obtained from structure in Figure 6.11b by adding one atom per unit cell in such a way that it bridges the two atoms of a dimer on one side, and rebonds on the other side.

6.5.1.5 Discussion

The data in Table 6.1 show clearly that the density of dangling bonds at the Si(114) surface is, in fact, uncorrelated with the surface energy. The lowest number of dbs per area reported here is 4, and it corresponds to $n = 213$ and $\gamma = 90.43\,\text{meV}/\text{Å}^2$ at the DFT level. The optimum structure (Figure 6.11a), however, has twice as many dangling bonds but the surface energy is smaller, $88.77\,\text{meV}/\text{Å}^2$. Furthermore, for the same number of atoms in the supercell ($n = 216$) and the same dangling bond density ($8\,\text{db}/3a^2\sqrt{2}$), the different reconstructions obtained via global searches span an energy interval of at least $5\,\text{meV}/\text{Å}^2$. These findings constitute a clear example that the number of dangling bonds cannot be used as a criterion for selecting model reconstructions for Si(114); we expect this conclusion to hold for many other high-index semiconductor surfaces as well.

The HOEP surface energy and the DFT surface energy also show very little correlation, indicating that the transferability of the interaction model [25] for Si(114) is not as good as, for instance, in the case of Si(001) and Si(105) [44]. The most that can be asked from this model potential [25] is that the observed reconstruction [43] is among the lower-lying energetic configurations – which, in this case, it is. We have also tested the transferability of HOEP for the case of Si(113), and found that, although the adatom interstitial models [50] are not the most stable structures, they are still retrieved by HOEP as local minima of the surface energy. We found that the low-index (but much more complex) Si(111)-7×7 reconstruction is also a local minimum of the HOEP interaction model, albeit with a very high surface energy. Other tests indicated that, while the transferability of HOEP to the Si(114) orientation is marginal in terms of sorting structural models by their surface energy, this potential [25] performs much better than the more popular interaction models [20,22], which sometimes do not retrieve the correct reconstructions even as local minima. Therefore, HOEP is very useful as a way to find different local minimum configurations for further optimization at the level of electronic structure calculations.

A practical issue that arises when carrying out the global searches for surface reconstructions is the two-dimensional periodicity of the computational slab. In general, if a periodic surface pattern has been observed, then the lengths and directions of the surface unit vectors may be determined accurately through experimental means (e.g., STM or LEED analysis): in those cases, the periodic vectors of the simulation slab should simply be chosen the same as the ones found in experiments. When the surface does not have two-dimensional periodicity, or when experimental data are difficult to analyze, then one should systematically test computational cells with periodic vectors that are integer multiples of the unit vectors of the bulk truncated surface, which are easily computed from knowledge of crystal structure and surface orientation. There is no preset criterion as to when the incremental testing of the size of the surface cell should be stopped – other than the limitation imposed by finite computational resources; nevertheless, this approach gives a systematic way of ranking the surface energies of slabs of different areas, and eventually finding the global minimum surface structure.

Motivated by a previous finding that larger unit cells can lead to models with very low surface energies (see, for instance, the example of Si(105) [44,45]), we have also performed global minimum search using GA for slabs of dimensions $6a \times a\sqrt{2}$, which correspond to a doubling of the unit cell in the $[22\bar{1}]$ direction. The ground-state structure at the HOEP level found in this case is still the corrugated model (Figure 6.11b) with a surface energy of $\gamma = 81.66\,\text{meV}/\text{Å}^2$. As a low-lying configuration, we again retrieve the original model [43] with $\gamma = 83.39\,\text{meV}/\text{Å}^2$. Furthermore, we also find several models that have surface energy in between the two values, characterized by the presence of different $3a \times a\sqrt{2}$ structures in the two halves of the $6a \times a\sqrt{2}$ simulation cell. This finding suggests that $[1\bar{1}0]$-oriented boundaries between different $3a \times a\sqrt{2}$ models on Si(114) are not energetically very costly: this is consistent with the experimental reports of Erwin et al., who indeed found (2×1) and $c(2 \times 2)$ structures next to one another [43].

6.5.1.6 Concluding Remarks

In conclusion, we have obtained and classified structural candidates for the Si(114) surface reconstructions using global optimization methods and density functional calculations. We have used both parallel tempering Monte Carlo procedure coupled with an exponential cooling [44] and the genetic algorithm [45]. Both of the methods are used in conjunction with the latest empirical potential for silicon [25], which has a better transferability in comparison with more popular potentials [20,22]. We have built a large database of structures (reported, in part, in Table 6.1) that were further optimized at the DFT level. The lowest-energy structure that we found (Figure 6.11a) after the DFT relaxation is the same as the one originally reported for Si(114)-2×1 in Ref. [43].

In addition, we have discovered several other types of structures (refer to Figures 6.11b and c, 6.12a, and 6.13a) that are separated (energetically) by $1-2\,\text{meV}/\text{Å}^2$ from the lowest-energy model [43]. Given that the relative surface energies at the DFT level have an error of $\pm 1\,\text{meV}/\text{Å}^2$ and that the experiments of Erwin et al. [43] already identified two reconstructions [(2×1) and $c(2 \times 2)$] whose surface energies are within $1-2\,\text{meV}/\text{Å}^2$ from one another, it is conceivable that some of the models in Figures 6.11–6.13 could also be found on the Si(114) surface. This prediction could be tested, for example, by high-resolution transmission electron microscopy experiments such as the ones reported recently for the Si(5 5 12) surface [51]. Low-energy electron diffraction experiments, as well as more STM measurements, could also shed light on whether there exist other structural models on a clean Si(114) surface than initially reported in Ref. [43].

6.5.2
Crystal Structure Prediction: Comparison between GA and MH

In this section, we revisit a couple of crystal structure prediction problems that were described or mentioned in Chapter 3. We have increased the system size to see how the search efficiency changes with respect to size of the problem. We find that the relative performance and underlying mechanism of genetic algorithms in crystal structure searches for Al$_x$Sc$_{1-x}$ strongly depend on the system composition as well as the size of the problem. Because of this strong dependence, caution should be taken in generalizing performance comparison from one problem to another even though they may appear to be similar. We also investigate the performance of GA for crystal structure prediction of boron with and without a priori knowledge of the lattice vectors. The results show that the degree of difficulty increases dramatically if the lattice vectors of the crystal are allowed to vary during the search. Comparison of GA with the MH algorithm (Section 6.3) at small (<10 atoms) to larger problem sizes is also carried out. At the small sizes we have tested, both methods show comparable efficiency, while for large sizes the GA performance becomes faster than that of MH.

6.5.2.1 GA Applied to Al$_x$Sc$_{1-x}$ Alloys

Our first benchmark study is on the Al$_x$Sc$_{1-x}$ alloy system. As mentioned in Chapter 3, this system has been previously investigated by Trimarchi and Zunger using a GA global optimization procedure similar to ours [9]. Their GA global search for lowest-energy crystal structures had been performed for 2–6, 4–4, and 6–2 atomic compositions and a total of eight atoms per unit cell are included in their calculations. We first performed a similar search with an eight-atom supercell and a pool size of 20 using our algorithm. The k-point mesh resolution is chosen as $2\pi \times 0.05$ Å^{-1} for the structure relaxations and twice as fine for the final energy calculations. The plane-wave cutoff energy is 241 eV. Our calculations yielded similar results as previous work [9]. We also obtained the D0$_{19}$ structure with space group $P6_3/mmc$ for Al$_2$Sc$_6$. For Al$_4$Sc$_4$, in addition to the B2 structure with space group $Pm3m$ found by Trimarchi and Zunger, an L1$_0$ structure with space group $P4/mmm$ is also identified as a low-energy structure. These are basically tetragonal distortions of the B2 structure (Figure 6.15). The energy differences between these different structures are smaller than 10 meV per eight-atom supercell.

The convergence behavior of Al$_2$Sc$_6$ and Al$_4$Sc$_4$ from the present GA search is similar to that reported in the work of Trimarchi and Zunger [9]. In Figure 6.14, the search histories for a number of runs for Al$_2$Sc$_6$ and Al$_4$Sc$_4$ are shown. As one can see from Figure 6.14a, in Al$_2$Sc$_6$ there are several energy drops in the GA evolution before the lowest-energy structure is reached, a typical behavior in GA search, but it takes only about 15 generations for the system to reach the ground-state configuration. For Al$_4$Sc$_4$, the ground-state structure is found almost immediately in several generations. For example, one of the runs in Figure 6.14b produces the ground-state structure even in the pool initialization step. Considering the small pool size, we think that for Al–Sc in this size range, a search using randomly generated initial structures already gives good structural guess without using the mating aspect of the genetic algorithm. The two nearly degenerate stable structures for Al–Sc, L1$_0$ and B2, are both cubic-based structures (Figure 6.15) that could make it easier for random initialization to locate the same low-energy structures by providing a larger capture basin. These results indicate that when the unit cell size is too small, the ground-state structures can be obtained from structure relaxation on randomly

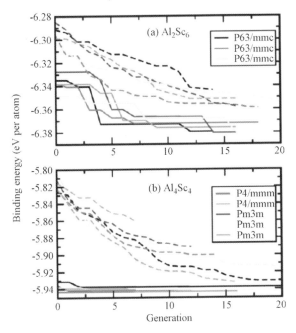

Figure 6.14 The history of lowest energy (per atom) of Al$_2$Sc$_6$ and Al$_4$Sc$_4$ by generation. Different curves represent each independent run and solid line represents the lowest-energy structure in every generation and dashed line the average pool energy. (From Ref. [42], with permission from Royal Society of Chemistry.)

generated initial structures and the efficiency of the genetic algorithm cannot be properly evaluated in such cases.

In order to test the GA for larger systems, we repeated the searches for the same Al$_x$Sc$_{1-x}$ alloys but with a double-sized supercell containing 16 atoms. The pool size is the same as in the search for small unit cells. These larger unit cell searches are more time consuming: doubling the system size requires much longer DFT calculation for the relaxation step. We report the results from five independent search runs for both Al$_4$Sc$_{12}$ and Al$_8$Sc$_8$.

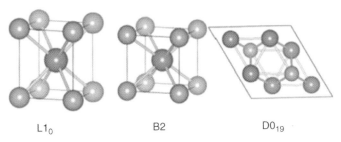

Figure 6.15 Structures of L1$_0$, B2 (AlSc), and D0$_{19}$ (AlSc$_3$). (From Ref. [42], with permission from Royal Society of Chemistry.)

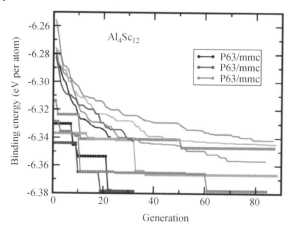

Figure 6.16 The history of lowest energy (per atom) of Al$_4$Sc$_{12}$ by generation. The final structure is identified as D0$_{19}$ with *P63/mmc* space group. Different curves represent each independent run and dotted line represents the lowest-energy structure in every generation and solid line the average pool energy. (From Ref. [42], with permission from Royal Society of Chemistry.)

As shown in Figures 6.16 and 6.17, much more generations and hence more trial structures need to be examined before the lowest-energy structure with 16 atoms per unit cell is found. For Al$_4$Sc$_{12}$, three in five GA runs give the ground-state structure (D0$_{19}$) and four in five for Al$_8$Sc$_8$ (B2). In general, the stepwise energy drop behavior is clearly seen in the GA evolution of the larger supercell structures for both Al$_4$Sc$_{12}$ and Al$_8$Sc$_8$. It is interesting to note that the GA evolutionary behavior of the 1 : 1 ratio Al$_8$Sc$_8$ is quite different from that of the small supercell Al$_4$Sc$_4$. In the small unit cell

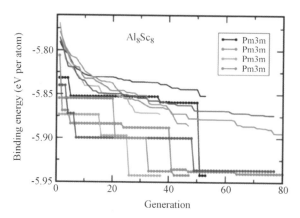

Figure 6.17 The history of lowest energy (per atom) of Al$_8$Sc$_8$ by generation. The final structure is identified as B2 with *Pm3m* space group. Different curves represent each independent run and dotted line represents the lowest-energy structure in every generation and solid line the average pool energy. (From Ref. [42], with permission from Royal Society of Chemistry.)

Figure 6.18 Six best structures in the final pool of Al_2Sc_6. Lattice vectors are aligned to highlight the structure similarity. (From Ref. [42], with permission from Royal Society of Chemistry.)

Al_4Sc_4, searches produce structures with energy very close to the ground-state structures within just a few generations. However, in the doubled supercell the GA search of 1 : 1 and 1 : 3 Al–Sc ratios shows similar convergence rate but with more generations (20–60) than their small supercell counterparts. To fully understand the GA behavior, we plot the six best structures in the final pool from one successful run for Al_2Sc_6 and Al_4Sc_{12}, respectively, as shown in Figures 6.18 and 6.19. All the structures are properly oriented and the unit cell is outlined to highlight similarities between structures. We see that for Al_2Sc_6 the structures in the final pool are very different. Contrary to the random character of Al_2Sc_6, in the large supercell search the low-energy structures of Al_4Sc_{12} in the pool share similar lattice types with different atomistic arrangement. The first structure in Figure 6.19 is the ground-state structure (DO_{19}), which is an hcp-based structure. In the other low-energy structures, the hcp stacking motif can be easily identified in different parts of the supercell. The real-space cut-and-paste GA hybridizes local motifs of parents to produce offspring and propagate better structure properties to new generations.

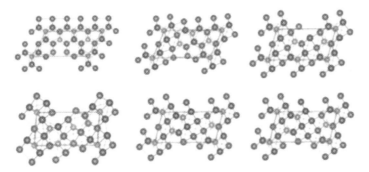

Figure 6.19 Six best structures in the final pool of Al_4Sc_{12}. Lattice vectors are aligned to highlight the structure similarity. (From Ref. [42], with permission from Royal Society of Chemistry.)

However, if the system size is too small and the length scale of the structure motifs from the parent structures is already larger than the system size, the cut-and-paste operation does not work effectively. That may explain why structures in the pool of Al_2Sc_6 are so different. The GA mating operation produces structures more like a random search in an extremely small system. A similar situation is also observed for Al_4Sc_4 versus Al_8Sc_8. Fortunately, for such small sizes, local structure optimization algorithms such as conjugate gradient can already locate the ground state in a small number of randomly chosen trials.

6.5.2.2 Boron

Recently, a new crystal structure of γ-B_{28} has been identified by a combination of experiment and computational crystal structure prediction using an evolutionary algorithm [10]. It will be interesting to benchmark our algorithm in this system where the number of atoms in the unit cell is large and the chemical bonding is also more complicated. We first performed the search with the number of boron atoms in the unit cell ranging from 24 to 32 and the lattice vectors fixed to the ones reported in Oganov et al.'s work [10]. The number of structures kept in the pool during the GA search is 64. The plane-wave energy cutoff is 318 eV and $4 \times 4 \times 4$ Monkhorst–Pack k-points are used. The lowest energies found as a function of boron number for fixed unit cell are plotted in Figure 6.20. Our GA search found that the B_{28} structure reported by Oganov et al. [10] has the lowest energy among all the structures from our search with system sizes ranging from 24 to 32 atoms per supercell. The binding energies per atom are consistent with the values previously reported [10]. We found that this B_{28} structure can be reached readily in about 15 GA generations. This performance is quite surprising considering the large number of atoms in the supercell and the complicated bonding nature of boron.

Note that in the GA search discussed above, the lattice vectors of the crystal structure are obtained from the X-ray experiment and are fixed during the GA search. It would be

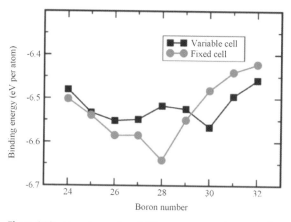

Figure 6.20 Lowest energies of different atom number in the final GA pool of B_{24-32} GA search with fixed and variable cell. (From Ref. [42], with permission from Royal Society of Chemistry.)

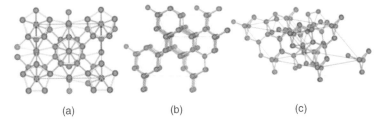

Figure 6.21 (a) Perfect γ-B_{28}. (b) Metastable structure of B_{28} found by GA search with fixed system volume but variable lattice vectors. (c) B_{30} found by GA search with fixed system volume but variable lattice vectors. Honeycomb structure combined with some B_{12} icosahedra can be seen. (From Ref. [42], with permission from Royal Society of Chemistry.)

interesting to see the efficiency of the GA search for a general case where the lattice vectors are not known a priori. In this search, the lattice vectors are varied with the system volume fixed at the same value as in our previous search. The atom number also varies from 24 to 32. The k-point mesh resolution is chosen as $2\pi \times 0.1 \text{Å}^{-1}$ for the DFT structural relaxations and twice as fine for the final total energy calculations. The binding energies (per atom) of the lowest-energy structures obtained from our GA search after 180 generations and nearly 3000 structure relaxations have been examined for each size ranging from 24 to 32 atoms and plotted as the black line in Figure 6.20. Comparing with the fixed-cell GA search, the result for B_{28} has much higher energy than that of the ground-state structure discussed above, even though many more GA generations have been performed in this new search. The energy of this variable-cell B_{28} structure is also higher than those of other structures with different number of atoms, although the volume of the unit cell should favor 28 atoms. These results indicate that without the knowledge of lattice vectors the GA search becomes much more difficult. The results also suggest that at 28 atoms the system is trapped in a local minimum structure with a much broader capture basin than that of the ground-state structure. As shown in Figure 6.21b, the metastable structure of B_{28} found in this search is very well ordered with sixfold honeycomb-like structures aligned perfectly in one direction. For comparison, the lowest-energy structure with 30 boron atoms obtained from this GA search is also shown in Figure 6.21c. Interestingly, B_{30} is the lowest-energy structure found in the variable-cell search and also energetically better than the one with the same cell size found in fixed-cell search. The B_{30} structure is a mixture of a honeycomb and B_{12} icosahedra units, which is the main feature of ground-state γ-B_{28}. This B_{30} structure is energetically better than the honeycomb-like metastable B_{28}. We have repeated the variable-cell-shape 28-atom search 10 times and all runs ended up trapped in the honeycomb-like metastable structures shown in Figure 6.21b.

From the test cases of boron, allowing lattice vectors to vary dramatically reduces the efficiency of the GA performance. We observed that in the GA search for the Al_xSc_{1-x} crystals, the lattice vectors converge very quickly and then the problem becomes nearly a fixed-cell problem in later generations. However, in the case of boron it is very hard to converge to the correct lattice vectors. Another point to note is that in our present variable-cell GA search, the volume of the unit cell is kept constant. This constraint of fixed volume might have an effect on the landscape for the structure search. We are pursuing further studies where the volume of the system is relaxed in the variable-cell studies or with the atom density kept constant in the fixed-cell searches. We hope the difference in behavior between these different ways of carrying out the search would throw some light on what is the best way to search for the global minimum structure when the unit cell vectors are not known from experiments. We also note that this behavior is similar to that observed in GA optimization of the C_{20} cluster described in our original paper [52]. In that case, we showed that the problem can be solved by including mutations or by running multiple ecologies. The method of atom-weighted average of two parent lattices (described in Chapter 3) that we currently used might not be the best method to use in GA schemes. At this point, how to best include lattice vector variations into the GA crossover operations is still an open question.

6.5.2.3 Minima Hopping

We have also performed some of the above crystal structure prediction using the MH algorithm. $AlSc_3$, AlSc with 8 and 16 atoms, and B_{28} have been tested. As mentioned in Section 6.3, in the MH approach, the initial structure is relaxed to a local minimum and then molecular dynamic simulation at elevated temperature is performed to help the system jump out of the current local minimum to another nearby potential minima. Another geometry relaxation is then performed to determine if the new local minimum is acceptable. By following the MH algorithm, the system travels around the potential energy surface and finally locates the global minimum. In the MH approach, the MD temperature and acceptance/rejection odds for the MH are parameters that can be adjusted to influence the performance of the algorithm. We adopt the same standard parameters proposed by Goedecker [7]: $\alpha_1 = 1/1.05$, $\alpha_2 = 1/1.05$, $\beta_1 = \beta_2 = 1.05$, and $\beta_3 = 1/1.05$. $\alpha_{1,2}$ determine how rapidly the energy criteria for rejection and acceptance change when a new configuration is being located. $\beta_{1,2,3}$ determine how rapidly the kinetic energy changes in the MD hopping process.

The plane-wave energy cutoff, k-point sampling, and so on in the first-principles calculations for structural relaxations during the MH searches are similar to those used in the GA search discussed above. The evolution of the Al_2Sc_6 and Al_4Sc_4 during the MH with various starting configurations is shown in Figures 6.22 and 6.23, respectively. In three out of eight runs, Al_2Sc_6 reaches its $D0_{19}$ ground-state structure within 200 MH steps. Almost all of the runs for Al_2Sc_6 can finally reach $L1_0$ or B2. However, for more challenging 16-atom supercell cases such as Al_4Sc_{12} and Al_8Sc_8, MH becomes quite inefficient as shown in Figures 6.24 and 6.25. In 10 runs for each composition, only one run finds the ground-state structure within 300 MH steps. In comparison with average trial structure number in GA (Figures 6.16 and 6.17), at large size GA will take advantage over MH algorithm.

For B_{28}, we have done the MH search with both fixed cell and variable cell shape, as we did in the GA search. In Figure 6.26, we show the evolution of B_{28} during the MH optimization with seven independent runs from different starting structures for both fixed-lattice and variable-lattice searches. In most cases, we can obtain the ground-state B_{28} structure within 150 MH steps in fixed-cell search. In one case, we can even reach the ground-state structure in tens of MH steps, as one can see from in Figure 6.26. This efficiency for the fixed-cell case is similar to that in the GA

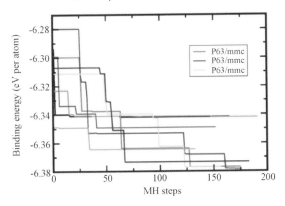

Figure 6.22 The energetic history of Al_2Sc_6 at every MH step for each individual run with different starting structures. Three of the runs eventually reached the DO_{19} ($P63/mmc$) structure. (From Ref. [42], with permission from Royal Society of Chemistry.)

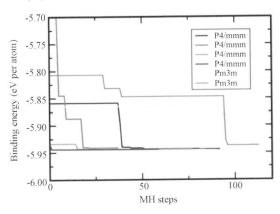

Figure 6.23 The energetic history of Al_4Sc_4 at every MH step for each individual run with different starting structures. Four of the six runs give the $L1_0$ ($P4/mmm$) structure and two give B2 ($Pm3m$). (From Ref. [42], with permission from Royal Society of Chemistry.)

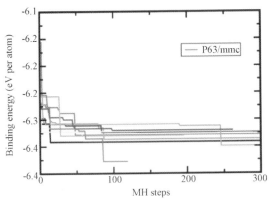

Figure 6.24 The energetic history of Al_4Sc_{12} at every MH step for each individual run with different starting structures. Only 1 run in 10 finally gives the DO_{19} ($P63/mmc$) structure. (From Ref. [42], with permission from Royal Society of Chemistry.)

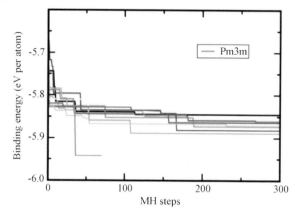

Figure 6.25 The energetic history of Al_8Sc_8 at every MH step for each individual run with different starting structures. Only 1 run in 10 finally gives the B2 (*Pm3m*) structure. (From Ref. [42], with permission from Royal Society of Chemistry.)

search. However, for variable-cell-shape search the systems are trapped in the metastable honeycomb-like tube structure as in the GA search.

We would like to point out that when first-principles total energy calculations are used in GA or MH for crystal structure predictions, the majority of the computer time is spent in the first-principles calculations. In our present study, each MH step consists of 100 MD steps followed by a relaxation step using the conjugate gradient method to find a local minimum, while the number of the trial structures that need to be relaxed in each GA generation is equal to the number of offspring (1/4 of the structures in the pool in our present study). Based on our current search statistics for AlSc and boron systems, in order to obtain a ground-state structure, the total number of MH steps is about 1.5 times greater than the total number of trial structures in the GA search. For example, in order to get the ground-state structure of Al_2Sc_6 GA it took about 120 MH steps in the MH search but 80 trial structure calculations (12 generations) in the GA

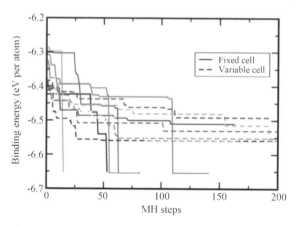

Figure 6.26 MH step history of B_{28} at fixed (solid line) and variable (dashed line) cell parameters. (From Ref. [42], with permission from Royal Society of Chemistry.)

search. However, we found that in the GA searches each trial structure needs more time for the relaxation by first-principles calculations because some far-from-equilibrium atomic configurations can be produced in the GA mating process. In contrast, the relaxation step in the MH search took much less time due to the continuity of the atomic trajectories in the MD simulations. Even with the overhead of 100 MD steps (which can be run with lower energy cutoff) before each relaxation, the total computer time spent on each MH step is only about 2/3 of the time used in the GA relaxation of one trial structure. For the case of Al_2Sc_6 studied in this chapter, it took about 50 min for each MH step, but needed over 80 min to relax one trial structure in the GA search. Based on above rough estimation, the computer time spent to find the same ground-state structure for small-size system by the both methods is similar. However, for larger-size systems with 16 atoms, most of the searches we have performed do not converge in 300 MH steps and even more. Thus, the MH method does not work as efficiently for large systems as GA. This performance difference at different problem sizes is consistent with the random effect that we have discussed for small systems. Only beyond certain size GA will begin to show its capability to generate better fitting structures and become more efficient than MH.

In summary, we have reported investigations of the performance of the genetic algorithm and the MH algorithms in global crystal structure optimization for the cases of Al_xSc_{1-x} metallic alloys and boron. For Al_xSc_{1-x} at small size, both GA and MH can find the ground-state structures easily. We found two stable structures $L1_0$ and B2 for Al_4Sc_4 with almost the same binding energy. If we change the size in the GA search from 8 atoms to 16 atoms for the Al_xSc_{1-x} alloys with the same compositions, the crystal structure prediction becomes more difficult as expected but the structures still can be predicted correctly even when the lattice vectors are not known a priori. However, the evolution histories are quite different between small- and large-size cases. For small unit cells, the current mating scheme just generates random-like offspring. For larger 16-atom supercell search, our GA mating operation shows up and GA search takes advantage over MH. In the case of γ-B_{28}, both GA and MH can easily locate the lowest-energy structure if the lattice vectors of the unit cell are given. However, when the lattice vectors are not known a priori, the problem becomes much more difficult. Therefore, search for the ground-state structure of B_{28} without knowledge of the lattice vectors presents a challenge for GA development. With lattice information, the structure search is not quite challenging enough for GA at the system size that can be handled by first-principles calculations at the present time.

References

1 Oganov, A.R. (ed.) (2010) *Modern Methods of Crystal Structure Prediction*, Wiley-VCH Verlag GmbH, Weinheim.
2 Kirkpatrick, S., Gellat, C.D., and Vechi, M.P. (1983) *Science*, **220**, 671.
3 Freeman, C.M., Newsam, J.M., Levine, S.M., and Catlow, C.R.A. (1993) *J. Mater. Chem.*, **3**, 531.
4 Schmidt, M.U. and Englert, U. (1996) *Dalton Trans.*, **10**, 2077.
5 Schön, J.C. and Jansen, M. (1996) *Angew. Chem., Int. Ed. Engl.*, **35**, 1287.
6 Wales, D. and Doye, J. (1997) *J. Phys. Chem. A*, **101**, 5111.
7 Goedecker, S. (2004) *J. Chem. Phys.*, **120**, 9911.

8 Xiang, H.J., Da Silva, J.L.F., Branz, H.M., and Wei, S.H. (2009) *Phys. Rev. Lett.*, **103**, 116101.
9 Trimarchi, G. and Zunger, A. (2008) *J. Phys.: Condens. Matter*, **20**, 295212.
10 Oganov, A.R., Chen, J.H., Gatti, C., Ma, Y.Z., Ma, Y.M., Glass, C.W., Liu, Z.X., Yu, T., Kurakevych, O.O., and Solozhenko, V.L. (2009) *Nature*, **457**, 863.
11 Stillinger, F.H. and Weber, T.A. (1983) *Phys. Rev. A*, **28**, 2408.
12 Stillinger, F.H. (1999) *Phys. Rev. E*, **59**, 48.
13 Wales, D.J. and Doye, J.P.K. (1997) *J. Phys. Chem. A*, **101**, 5111.
14 Voter, A.F. (1998) *Phys. Rev. B*, **57**, R13985.
15 Sørensen, M.R. and Voter, A.F. (2000) *J. Chem. Phys.*, **112**, 9599.
16 Geyer, C.J. and Thompson, E.A. (1995) *J. Am. Stat. Assoc.*, **90**, 909.
17 Hukushima, K. and Nemoto, K. (1996) *J. Phys. Soc. Jpn.*, **65**, 1604.
18 Kirkpatrick, S. (1984) *J. Stat. Phys.*, **34**, 975.
19 Balamane, H., Halicioglu, T., and Tiller, W.A. (1992) *Phys. Rev. B*, **46**, 2250.
20 Tersoff, J. (1988) *Phys. Rev. B*, **38**, 9902; Tersoff, J. (1988) *Phys. Rev. B*, **37**, 6991.
21 Nurminen, L., Tavazza, F., Landau, D.P., Kuronen, A., and Kaski, K. (2003) *Phys. Rev. B*, **67**, 035405.
22 Stillinger, F.H. and Weber, T.A. (1985) *Phys. Rev. B*, **31**, 5262.
23 Chelikowsky, J.R., and Phillips, J.C., (1990) *Phys. Rev. B*, **41**, 5735; Chelikowsky, J.R., Glassford, K.M., and Phillips, J.C. (1991) *Phys. Rev. B*, **44**, 1538.
24 Bolding, B.C. and Andersen, H.C. (1990) *Phys. Rev. B*, **41**, 10568.
25 Lenosky, T.J., Sadigh, B., Alonso, E., Bulatov, V.V., Diaz de la Rubia, T., Kim, J., Voter, A.F., and Kress, J.D. (2000) *Model. Simulat. Mater. Sci. Eng.*, **8**, 825.
26 Metropolis, N., Rosenbluth, A.W., Rosenbluth, M.N., Teller, A.M., and Teller, E. (1953) *J. Chem. Phys.*, **21**, 1087.
27 Hansmann, U.H.E. (1997) *Chem. Phys. Lett.*, **281**, 140.
28 Geman, S. and Geman, D. (1984) *IEEE Trans. Pattern Anal. Mach. Intell.*, **6**, 721.
29 Hajek, B. (1988) *Math. Oper. Res.*, **13**, 311.
30 Sugita, Y., Kitao, A., and Okamoto, Y. (2000) *J. Chem. Phys.*, **113**, 6042.
31 Predescu, C., Predescu, M., and Ciobanu, C.V. (2004) *J. Chem. Phys.*, **120**, 4119.
32 Moreno, J.J., Katzgraber, H.G., and Hartmann, A.K. (2003) *Int. J. Mod. Phys. C*, **14**, 285.
33 Matsumoto, M. and Nishimura, T. (2000) in *Monte Carlo and Quasi-Monte Carlo Methods 1998* (eds H. Niederreiter and J. Spanier), Springer, New York, pp. 56–69.
34 Amsler, M. and Goedecker, S. (2010) *J. Chem. Phys.*, **133**, 224104.
35 Ghasemi, A., Goedecker, S., Lenosky, T., Hug, H., Meier, E., and Baratoff, A. (2008) *Phys. Rev. Lett.*, **100**, 236106.
36 Roy, S., Goedecker, S., Field, M.J., and Penev, E. (2009) *J. Phys. Chem. B*, **113**, 7315.
37 Bao, K., Geodecker, S., Genovese, L., Neelov, A., Ghasemi, S.A., and Deutsch, T. arXiv:00902.1599.
38 Andersen, H.C. (1980) *J. Chem. Phys.*, **72**, 2384.
39 Parrinello, M. and Rahman, A. (1980) *Phys. Rev. Lett.*, **45**, 1196; Parrinello, M. and Rahman, A. (1981) *J. Appl. Phys.*, **52**, 7182.
40 Martoňák, R., Laio, A., and Parrinello, M. (2003) *Phys. Rev. Lett.*, **90**, 075503.
41 Chuang, F.C., Ciobanu, C.V., Predescu, C., Wang, C.Z., and Ho, K.M. (2005) *Surf. Sci.*, **578**, 183.
42 Ji, M., Wang, C-Z., and Ho, K.-M. (2010) *Phys. Chem. Chem. Phys.*, **12**, 11617.
43 Erwin, S.C., Baski, A.A., and Whitman, L.J. (1996) *Phys. Rev. Lett.*, **77**, 687.
44 Ciobanu, C.V. and Predescu, C. (2004) *Phys. Rev. B*, **70**, 085321.
45 Chuang, F.C., Ciobanu, C.V., Shenoy, V.B., Wang, C.Z., and Ho, K.M. (2004) *Surf. Sci.*, **573**, L375.
46 Baski, A.A., Erwin, S.C., and Whitman, L.J. (1995) *Science*, **269**, 1556.
47 Mo, Y.W., Savage, D.E., Swartzentruber, B.S., and Lagally, M.G. (1990) *Phys. Rev. Lett.*, **65**, 1020.
48 Baroni, S., Dal Corso, A., de Gironcoli, S., and Giannozzi, P. http://www.pwscf.org.
49 Perdew, J.P. and Zunger, A. (1981) *Phys. Rev. B*, **23**, 5048.
50 Dabrowski, J., Müssig, H.J., and Wolff, G. (1994) *Phys. Rev. Lett.*, **73**, 1660.
51 Liu, J., Takeguchi, M., Tanaka, M., Yasuda, H., and Furuya, K. (2001) *J. Electron Microsc.*, **50**, 541.
52 Deaven, D.M. and Ho, K.M. (1995) *Phys. Rev. Lett.*, **75**, 288.

7
Perspectives and Outlook

7.1
Expansion through the Community

At present, the genetic algorithm (GA) applications to crystal structures have become so advanced that USPEX (mentioned in Chapter 1) and perhaps other codes have become available to potential users worldwide. This will probably sustain scientific interest in using GA for atomic structure problems for a while, and, furthermore, will also ensure a steady supply of developers. Development is still necessary given that compounds and materials with large unit cells and large numbers of atomic species in the unit cell are likely beyond reach with available codes. The current routine capability of the GA is at approximately 100 atoms per unit cell. Doubling this number by discovering more efficiency in the programming or implementation of GA will ensure access to wider range of problems, with increasingly larger technological relevance. For example, the current GA codes cannot yet tackle the situations in which there are impurities of many kinds (and tiny amounts) in an otherwise known crystal structure because that implies too large unit cells. GA is yet to be able to reproduce high-grade industrial steels for precise applications, steels that have many atomic species "ingredients" and highly complex phase diagrams.

7.2
Future Algorithm Developments

In Chapter 6, we described a few methods used for global minimum searches and, after performing comparisons with GA we surmised that GA was the faster method. While that may be true for sufficiently large number of atoms in the simulation cell, the comparison with parallel tempering Monte Carlo (PTMC) annealing was not quite "fair" because this method is designed to reproduce the materials thermodynamics, whereas GA is not. In effect, the fundamental scope of the PTMC method is different from that of GA, although in Chapter 6 they are used for the very same purpose (global minimum search). PTMC deals with the full potential energy surface (PES), while GA deals with the reduced PES. The PTMC method carefully samples the barriers in the potential energy landscape, while GA does not. This brings us to the issues of "control" variable, a variable that can be adjusted dynamically during a run in such a way as to ensure continuous progress. In PTMC, this is the maximal displacement of any given atom, which is adjusted to keep the acceptance rate of the individual atomic moves at about 50%. In basin hopping Monte Carlo, which deals with the reduced PES just like GA, there exists a similar control variable. In GA, on the other hand, there is no such variable that controls a steady acceptance/rejection rate for the crossovers. As a result, when a GA run is started from random structures, it always progresses very fast in the beginning, after which it quickly slows down to the point that acceptance of a new structure becomes a matter of luck. This is a fundamental area where developments should probably occur in order to deal with increasingly larger systems.

In addition to the control variable, another area that can benefit from additional developments in order to tackle new and specific structure determination problems would be the representation of the material and the crossover operations. For example, if we know that we are looking for a material that can be understood as a set of tetrahedra sharing corners, there is little to be gained by allowing the cut-and-paste operation to cut those individual tetrahedra. In this case, the material should be represented by tetrahedra (instead of atoms) and the GA operations should be changed accordingly. Similar ideas are for molecular crystalline materials, where the molecules that make up the material should be preserved by the GA operation. Another key improvement regarding crossovers has already been pointed out in Chapter 6: in the GA for crystal structure prediction, at present it is not clear that the assignment of lattice vectors to a new, child structure as a weighted average of the lattice vectors of the parents is a good way to pass on structural motifs of the parents to the child. Despite the current successes of GA, the assignment of the simulation box to a child structure may be significantly improved, which in turn will also improve the overall performance of the algorithm.

7.3
Problems to Tackle – Discovery versus Design

In Section 2.1.4, we have briefly summarized the range of structural optimization problems that GA can tackle. So far most of the progress occurred for cluster and crystal structure optimizations. The algorithm developments for surface, interfaces, and nanowires are done and published (see details in Chapters 1 and 5), but the applications so far are not as numerous as those for crystals or clusters. With increasingly faster computational resources becoming available, we envision that applications to structural problems that require "templates" (i.e., those in which the numbers of atoms subjected to GA operations are only a small subset of the atoms required to relax in order to compute energetic cost functions) will increase significantly. For example, general problems that can benefit from application of GA are determining the structure of a dislocation core, finding the structure of clusters embedded in a crystalline matrix, or finding new structures of grain boundaries in diverse materials – problems that are important in many areas of materials science, physics, and mechanical engineering. Other type of problems to tackle may be the discovery of framework materials, molecular crystals, or protein crystals, as already mentioned. In addition, there may be technologically motivated problems that do not rely on crystalline structures: for example, producing the atomic structure of an amorphous material or compound for which some macroscopic parameters are known (say, the density or the amount of hydrogen etc.) can also be of great interest since computer simulations studies of these materials require us to deal with the correct structure first.

We note that throughout this book GA was used to either reproduce well-known structures (benchmark) or to discover new structures once the benchmarks have been achieved. While discovery is fundamentally important, it still does not quite amount to materials design, that is, to creating crystal structures or nanomaterials with desired properties. As described in this book, GA leads to ground-state structures, polymorphs, and a number of metastable states and operated on energy-based selection criteria. The properties (electronic, magnetic, optical, mechanical, etc.) of these low-energy or low-enthalpy states can be computed after the GA run; most likely, after such assessment of the properties we are likely to discover that the desired properties are not associated with the ground state but with a metastable state. But how does one go about producing or synthesizing this metastable state with desired properties? Is it even possible to synthesize such a state? If not, then what material can be obtained that still has properties close enough to the desired ones? Questions such as these are currently on the table of computational materials designers. One can envision, for example, changing the cost function in the GA implementation from an energy-based quantity to the property for which a material has to be designed. This is an important and seemingly sensible approach, only the main impediment is that the calculation of the design property to be used as fitness function in a GA implementation is usually much more expensive computationally than the calculation of energy or enthalpy. Even if new code developments make computation of design properties faster, the problem of whether or not the material can be eventually synthesized still remains. Its solution will require close interaction with experimentalists that would lead to simulation and optimization of processes and it is likely that GA approaches will be instrumental there as well.

Index

a

ab initio
- calculations 6, 125, 153
- density functional methods 100, 165
- packages 6
- random structure searching method 63
- relaxations 107

AB Lennard-Jones mixtures 163
AIRSS. *See ab initio,* random structure searching method
algorithms
- crossover 133–135
- genetic 130, 132
- ingredients 135
- Metropolis 153
- for nanotubes 131
- particle swarm 132
- selection 133
- for SiNW optimization 125

Al_2Sc_6
- lowest energy, history of 177
- six best structures 179

Al_4Sc_{12}, best structures 179
Al_4Sc_4 lowest energy, history of 177
Al–Sc system convex hull 50
atomic clusters optimization 20, 21, 71–84
- alloys, oxides, and cluster materials 71–73
- GA solution to Thomson problem 81–84
- substrate-supported clusters optimization via GA 73–81

atomic configurations, evolution of 164
atomic interaction model 41
atomic structure prediction
- challenge 1–7
- evolution, reality and algorithms 2, 3
- historical perspective 4–6
- scope and organization 6, 7

atomic structure search strategy 74
average DFT pressure
- convergence 59

b

basin hopping Monte Carlo (BHMC) 150, 158–160
Bell–Evans–Polanyi principle 161, 162
BHMC. *See* basin hopping Monte Carlo (BHMC)
binary representation 71
- crossover operations 26
bismuth nanolines 107
Boltzmann constant 153
Boltzmann distribution 25, 28
bound system 12
Brillouin zone 125, 168
- integration 59
buckyball structure 12, 13

c

carbon nanotubes (CNTs) 131, 136, 138, 144
carbon phases 18
Cartesian components 29
Cartesian coordinates 31
cation coordination number (CN) 51
chemical potentials 18, 19
Clapeyron slopes 55
cluster atoms 80
cluster structures 28
C_n clusters structure 14
coherent potential approximation 63
computational cell 40
conjugate gradient relaxation 81
constant-pressure variable-cell-shape molecular dynamics (MD) 52
cosine metric 46

Atomic Structure Prediction of Nanostructures, Clusters and Surfaces,
First Edition. Cristian V. Ciobanu, Cai-Zhuang Wang, and Kai-Ming Ho.
© 2013 Wiley-VCH Verlag GmbH & Co. KGaA. Published 2013 by Wiley-VCH Verlag GmbH & Co. KGaA.

cost function 49
cotunnite-type SiO_2
– structural parameters 53
crystal structure
– energy landscape complexity 38–40
– Fe–Co alloys, structure and magnetic properties 63–67
– GA and MH 175
– – Al_xSc_{1-x} alloy system 176–180
– – boron 180–182
– – minima hopping 182–185
– GA efficiency improvement 40, 41
– generation-zero structures creation 44, 45
– interaction models 41–44
– optimizations 189
– post-pyrite phase transitions identification 51–57
– prediction 22, 37–67
– structural diversity assessment of pool 45–48
– – fingerprint functions 45–47
– – PES general features 47, 48
– ultrahigh-pressure phases of ice 57–63
– units 117
– variable composition 48–50
CsCl crystal structures 2
cut-and-paste operations 27, 30
cut-and-splice operations 27

d
dangling bonds 19
degrees of freedom 40
degrees of success 34
density functional theory (DFT) 37, 41, 42, 87, 167
– band gaps 61
– calculations 13
– formation energies per Si atom 127
– relaxations 44, 99, 103, 170
– – structural 37
– surface energy 174
dimers 74, 97, 99, 102, 104, 106, 114, 171, 173

e
energy-and-force fitting procedure 43
Euler's formula application 83
exchange-correlation energy 168

f
face-centered cubic (fcc) structure 1
Fe–Co alloys
– magnetic properties 38
– structure and magnetic properties 63–67
– – computational details 63, 64

– – results and discussion 64–67
– structures 63
$Fe_{11}Co_5$ low-energy structures
– formation energies 65
Fe–Mg alloys 50
– convex hull 50
Fe_2P-to-cotunnite transition 56
Fe_2P-type SiO_2
– enthalpies 54
– structural parameters 53
fingerprint functions 45–47, 46
– components 46
first-principles
– calculations 2
– total energy calculations 76
fitness and sorting process 24
fitness function 24
force-matching method 58
formation energies 112
– per Si atom 129

g
GA. *See* genetic algorithm (GA)
GBs. *See* grain boundaries (GBs)
generalized gradient approximation (GGA) 55
generation zero genetic pool 23
generic methods 3
genetic algorithm (GA) 2, 11, 15, 30, 37, 44, 67, 71, 75
– applications 5, 38, 187
– crossover operations 26–30
– – crossovers and periodic boundary conditions 28–30
– – cut-and-slice crossover in real space 27, 28
– development 4, 48
– DFT optimization 124, 125
– discovery *versus* design 188, 189
– efficiency 40
– evolution, ingredients of algorithm 135 (*See also* algorithms)
– flowchart 42, 43
– general procedure 23, 24
– global search 78
– for grain boundary structure optimization 115, 116, 132
– interface structures 114, 115
– mutations 30–33
– – regular mutations 31–33
– – zero-penalty mutations 31
– nanowire and nanotube structures 123
– power 44
– processes 23
– progress 31

– properties 38
– prototype systems, useful conclusions 139–144
– – acceptance ratios for single-crossover 141, 143
– – equiprobable combination of crossovers 142–144
– – evolution of number of atoms 139, 140
– in real-space representation 11–34
– search 32
– selection of parent structures 24–26
– stopping criteria and subsequent analysis 34
– structural optimization techniques 165
– – density functional calculations 167–169
– – PTMC simulation 166, 167
– – Si(114), reconstructions of 165, 166
– structure determination problems 12–23
– – cluster structure 12–16
– – crystal structure prediction 16–19
– – range of applications 21–23
– – surface reconstructions 19–21
– structures generated by 116–120
– updating genetic pool, survival of fittest 33, 34
– usages 4
genetic operators 4
genetic pool 3, 11, 30, 33, 45
genetic sequence 82
germanium clusters 16
Gibbs free energies 51
global minimum energy configurations 6
global optimization methods 12
global optimizer 153
global search methods 20
global structural optimization 81
grain boundaries (GBs) 114, 115
– structural units for Si[001] 116
grain boundary energy 121
– calculations 121–123
– as a function of rotation angle 123
graphene 87
ground-state structure prediction methods 2

h

Hansel–Vogel (HV) potential 125
high-index semiconductors surfaces 20
high-index surfaces 88
– algorithm details 107
– atomic structure of steps 107
– supercell geometry 107
highly optimized empirical potential (HOEP) 93, 95–97, 101–103, 105, 153, 170, 174, 175

honeycombs (H) 102, 105, 181, 184
hopping techniques 82

i

ice
– ultrahigh-pressure phases 57–63
– – computational details 58, 59
– – crystal structures 60
– – electronic band gaps 62
– – enthalpies 61
– – results and discussion 59–63
interaction models 41–44
– *ab initio* methods 42
– adaptive classical potentials 42–44
– classical potentials 41
interface structures, genetic algorithm 114, 115

l

Lamarckian evolution 3
large-scale atomic/molecular massively parallel simulator (LAMMPS) code 59
LDA. *See* local density approximation (LDA)
LEED. *See* low-energy electron diffraction (LEED)
Lennard-Jones clusters
– nonicosahedral global minima of 159
– structure 73
Lennard-Jones (LJ) nanotubes 131
Lennard-Jones particles 133
Lennard-Jones potentials 58
Lennard-Jones systems 6
local density approximation (LDA) 52, 55
low-angle grain boundary 123
low-energy electron diffraction (LEED)
– analysis 174
– measurements 89
lowest-energy structure 84

m

magic clusters 73, 77, 79
magic nanowires 128
magic numbers 83
– structures 112
magnetic moments 65, 66
mass transport phenomena 74
mating pool 25
metadynamics approach 163, 164
metallic alloys
– clusters
– – ground-state structures 72
– global minima 72
metastable state 189
Metropolis logic 155

Mg–Fe systems 50
MgSiO$_3$
– post-perovskite (PPV) transition 51
MH algorithm 161
miniaturization 131
minima hopping (MH) 150
molecular dynamics (MD) simulations 150
Monkhorst–Pack sampling scheme 59
Monte Carlo simulation 73, 81, 82, 153, 154, 156, 157, 159, 160, 165
Monte Carlo walkers 150, 153, 157, 158
mutation operators 49

n

NaCl crystal 1
– structures 2
NaMgF$_3$ perovskite 57
nanoparticles 71
nanostructure 1
Nanowire structures 127
– predictions 132
NbCoB phase 53
NbCoB-type SiO$_2$
– enthalpies 54
– structural parameters 53

o

odd–even oscillations 126
one-dimensional nanostructure
– description of algorithm 132–135
– effectiveness of a given crossover 139–142
– equiprobable combination of crossovers 142–144
– evolution of number of atoms 139
– under radial confinement 130, 131
– results for prototype nanotubes 135–139
optimization, via Minima hopping 160–163
oxide nanoparticles 73

p

parallel tempering annealing method 5
parallel tempering Monte Carlo (PTMC) annealing 88, 89, 91, 149, 150, 151–153, 154, 155, 157, 165, 166, 187, 188
– algorithm as global optimizer 153, 154
– algorithm, description of 154–158
Parrinello–Rahman technique 163
passivated silicon nanowires 123–130
Perdew–Burke–Ernzerhof (PBE)-type generalized gradient approximation 59
periodic lattice vector 41
periodic simulation cell, general shape 17
PES. *See* potential energy surface (PES)

phonon
– calculations 55
– dispersions 62
– frequencies 52
– instabilities 56, 57
– optical 164
– related deformations 60
– and vibrational density of states 59
plate-like Ge clusters 16
potential energy 39, 133
potential energy landscape, schematic depiction 39
potential energy surface (PES) 3, 4, 24, 34, 47, 48, 149, 150, 159, 162, 187, 188
– characteristics 47
– funnel structure 48
– reduced 49
– theory 102
pressure-induced phase transitions 60
pressure–temperature phase diagram 56
projector-augmented wave (PAW) method 64
pyrite-type SiO$_2$ 51, 55, 56
– enthalpies 54

q

quantitative diversity 47
quantum-ESPRESSO package 59
quasi-entropy 47
quasiharmonic (QHA) free-energy calculations 51

r

random initial pool 44, 45
random selection 24, 25
rare-earth nanowires 107
real-space
– GA method 7
– genetic algorithms 21
– representation 29, 71
real-space cut-and-paste crossover operation 27
rebonded atoms(R) 102
reduced energy landscape 159
replica-exchange Monte Carlo method 153
roulette-wheel selection 25, 26

s

SA technique 150
scanning tunneling microscopy (STM) 74, 88, 162
semiconductor surfaces 19
Si(103)
– model reconstructions 98

– new reconstructions, for related surface 95–99
Si(105)
– computational details 89–91
– heat capacity of 155
– mating operation 90
– results 91–95
Si(114)
– azimuthal angle 109
– crystallographic analysis 108
– different reconstructions 169
– genetic algorithm, results 113
– low-energy structures 111
– magic-number structures 112
– step structures 110–114
– supercell dimensions 109
Si(337)
– model reconstructions for 99
– reconstructions 104, 105
– results for 101–105
Si cluster 81
– magic 74
silicon clusters 79
silicon nanowires (SiNWs) 5, 123, 124
simulated annealing (SA) 149
single-shot calculations 44
single-species cluster 22
sinusoidal functions 28
SiNWs. *See* silicon nanowires (SiNWs)
SiO_2
– lower-pressure phases 52
– phase diagram 55
Si(114)-(2×1) reconstruction
– density functional calculations 167–169
– discussion 174, 175
– GA results 167
– PTMC simulation 166, 167
– structural models 169–174
– surface energy 170, 171
Slater–Pauling curve 66, 67
small carbon clusters, structures 14
solar energy application devices 115
Sterling's approximation 39
Stillinger–Weber potential 121, 152
stopping criteria 34
structural motifs 24
surface energy 20, 92–94
– computed at DFT level 101
– minimization 106
– retrieved by genetic algorithm for Si 97
survival of the fittest 23, 33, 91
Swapping mechanism 153

t
TBMD. *See* tight-binding, molecular dynamics
temperature schedule 153
tetrahedral networks 60
tetramers 97, 102, 106, 112
tight-binding 13
– calculations 5
– molecular dynamics 76, 80
– potential 74, 75, 78
tournament selection 25
tricapped triangular prisms 53
two-body potential 47

u
ultrahigh-pressure ice
– lowest-enthalpy structures 58
– phase diagram 63

v
VASP code 64, 121
vibrational density of states (VDOS) calculations 59
virtual crystal approximation 63

w
Wulff construction 130

z
Z-contrast electron technique 120
zero-penalty mutations 30, 31, 45
zero-point motion (ZPM) effects 59